Mining and Minerals

A Practical Guide to the Life Cycle of a Mining Project

Consulting Editors **Darrell W Podowski** and **Jennifer Poirier**

Consulting editors
Darrell W Podowski and Jennifer Poirier

Managing director
Sian O'Neill

Mining and Minerals: A Practical Guide to the Life Cycle of a Mining Project
is published by

Globe Law and Business Ltd
3 Mylor Close
Horsell
Woking
Surrey GU21 4DD
United Kingdom
Tel: +44 20 3745 4770
www.globelawandbusiness.com

Printed and bound by CPI Group (UK) Ltd, Croydon CR0 4YY

Mining and Minerals: A Practical Guide to the Life Cycle of a Mining Project

ISBN 9781787421288
EPUB ISBN 9781787421295
Adobe PDF ISBN 9781787421301
Mobi ISBN 9781787421318

Table of contents

Introduction

Darrell W Podowski
Jennifer Poirier
Cassels Brock & Blackwell LLP

The exploration and exploitation of minerals is an international industry; year after year, mining projects are acquired, developed and decommissioned all around the globe. While each jurisdiction presents its own unique set of considerations, many essential concepts and principles are of general application, regardless of where the mining project is located, the commodity that is being mined and whether the project is on land, on ice or at the bottom of the sea. This practical guide outlines in a single volume the essential concepts and principles involved in the life of a mining project, from beginning to end.

Each chapter – written by leading practitioners in their particular fields around the world – describes concepts and principles relevant to the various stages of a mining project. Topics featured include key agreements – from confidentiality and preliminary agreements, option, earn-in and strategic alliance agreements to joint venture and shareholders' agreements; mining capital finance and project finance; sale and purchase of mining projects; and decommissioning of mining assets. In addition, this guide contains chapters focused on a discussion of Aboriginal peoples, social and environmental factors, as well as the international framework in which mining projects operate. Although the subject matter of a chapter may reflect the favoured practice in a particular country, the authors have attempted to include international applications of the concepts and principles described so that this guide can be used as a basis for application in multiple jurisdictions.

Certain chapters also summarise some of the latest trends that affect the mining industry, such as the ongoing development of requiring a social licence to conduct mining operations in various countries; the increasing need for interaction and consultation with Aboriginal peoples; and new types of alternative financing methods used to fund the acquisition and development of such projects when traditional financing sources (eg, capital markets and debt financing) may not be available.

The editors and authors of this guide have strived to provide a useful, informative and practical guide for anyone seeking a better understanding of key commercial and legal principles and documentation involved in finding, acquiring, developing, monetising and decommissioning a mining project.

Confidentiality and standstill agreements in the mining industry

Fred R Pletcher
Borden Ladner Gervais LLP

1. Introduction

Mining industry participants commonly seek to protect proprietary information regarding technical processes or procedures, information regarding the location of geologic prospects and technical data pertaining to the mineral characteristics of mining properties, all of which can be extremely valuable. Such protection is principally sought at the outset of negotiations for the sale or lease of mineral properties, between joint venture participants and partners engaged in development of mineral prospects, and in connection with the financing or acquisition of mining companies. Confidentiality and standstill agreements are typically employed in mining transactions to protect against the misuse of such non-public information, while facilitating the free flow of technical, economic and other data.

Care should be taken in negotiating and drafting these documents, as disputes respecting confidentiality and standstill arrangements in the mining industry can arise among even the most sophisticated and well-advised parties. These disputes are often noteworthy for their sheer financial magnitude, with hundreds of millions of dollars often at stake.

This chapter canvasses several noteworthy cases involving the use and misuse of confidential information in the mining industry and considers some of the principal issues in the negotiation and drafting of confidentiality and standstill agreements.[1] The paper draws on common law principles and case law from Canada and the United States, as those are the jurisdictions that have litigated these types of agreements in the mining industry the most. However, these principles and guidelines can be used in other jurisdictions after confirming with local counsel that they are consistent with local laws.

1 For other discussions of confidentiality agreements in the context of mining transactions, see Fred R Pletcher and Anthony A Zoobkoff, "Gotcha! Turning Confidentiality and Standstill Agreements Into Gold Mines" (2007) 53 *Rocky Mt Min L Inst* 28-1; Lynn P Hendrix, "Covenants Not to Compete and Confidentiality Agreements in the Natural Resources Industry" (1992) 38 *Rocky Mt Min L Inst* 21-1; Fred E Ferguson, Jr, "Confidentiality Agreements in the Mining Industry" (1989) 35 *Rocky Mt Min L Inst* 7-1; and Frank Erisman and John D McCarthy, "Obligations not to Use Confidential Information Disclosed During Negotiations to Acquire Interest in Mineral Properties: Observations on International Corona Resources Ltd v LAC Minerals Ltd" (1988) 33 *Rocky Mt Min L Inst* 23-1.

2. Common law principles regarding confidential information

2.1 Basis of claims for wrongful disclosure or misuse of confidential information

In the absence of a confidentiality agreement, Canadian and US courts have developed common law principles to address unlawful or wrongful use or misuse of confidential information.

A claim for wrongful disclosure or misuse of confidential information may be founded on the following common law and equitable causes of action:

- breach of an implied term of a contract;
- breach of confidence;
- breach of fiduciary duty;
- unjust enrichment; or
- wrongful interference with the contractual relations of others.[2]

Of the foregoing, the two most commonly relied upon causes of action for the improper use or disclosure of confidential information – not based on breach of contract – are breach of a general duty of confidence and breach of a fiduciary duty, both of which are discussed below

(a) Duty of confidence

Breach of confidence was first recognised as a distinct cause of action in Canada in *International Corona Resources Ltd v LAC Minerals Ltd.*[3] The Supreme Court of Canada held that LAC Minerals had misused confidential information which it had obtained during negotiations with International Corona in respect of a possible joint venture to secure for itself alone a commercial opportunity (ie, mineral rights adjacent to the international property) that LAC Minerals would otherwise have known nothing about. In defining the test for breach of the common law duty of confidence, the court relied on an earlier English Chancery decision, *Coco v Clark (Engineers) Ltd*, wherein Justice Megarry set out three requirements that must be satisfied by a plaintiff:

- The information transferred has the "necessary quality of confidence" about it;
- The circumstances of the transfer are such that the recipient is under an obligation to the confider with respect to the information; and
- The information has been used for a purpose other than that for which it was intended.[4]

---

2 Julie A Thorburn and KG Fairbairn, *Law of Confidential Business Information* (2006) at 5-2.
3 [1989] 2 SCR 574 at 635.
4 [1969] RPC 41 (Ch D) at 47.

The first branch of the test requires a disclosing party to establish that there has been an exchange of specific and identified information that meets the test of confidentiality. Generally, the test for confidentiality will be met by establishing that the information is not readily accessible to others.[5] In some circumstances, however, the courts will employ what has come to be known as the 'springboard' or 'long start' doctrine to establish whether the test of confidentiality has been met. Under this doctrine, the court determines whether the information in question has enabled the defendant to do something it could not otherwise have done as quickly without the information, even where sources of information may be known and available to the public. As was stated by Lord Green in *Saltman Engineering Co v Campbell Engineering Co*:

> *it is perfectly possible to have a confidential document, be it a formula, a plan, a sketch, or something of that kind, which is the result of the work done by the maker of materials which may be available for the use of anybody; but what makes it confidential is the fact that the maker of the document has used his brain and thus produced a result which can only be produced by somebody who goes through the same process.*[6]

The second branch of the test will be satisfied whenever information is communicated for a "limited purpose".[7] In other words, it must be shown that the confidant knew, or ought to have known, that the information was disclosed for a limited purpose and was not to be used for any purpose except that for which it was disclosed.[8]

Under the final branch of the test, the burden shifts to the receiving party, which must show that the information was used within the limited purpose for which it was intended. If this burden is not met, the final branch of the test will be satisfied. Industry practice or custom may be relevant to a determination of whether information has been misused. For example, in *Cadbury Schweppes Inc v FBI Foods Ltd*, the British Columbia Supreme Court held:

> *All parties to these proceedings agree that it is the custom in the food industry to recognize product-specific formula and process information as confidential and proprietary. Such information will not be disclosed to others or otherwise used, whether by employees, licensees, consultants, or others, without the prior knowledge and consent of the owner. This industry custom operates whether or not an express confidentiality agreement is in place, and applies even to entirely generic products.*[9]

---

5 *Id*. See also *Ridgewood Resources Ltd v Henuset (*1982), 35 AR 493 at 500 (CA); *Pharand Ski Corp v Alberta* (1991), 37 CPR (3d) 288 (Alta QB).

6 [1963] All ER 413 at 415 (CA). See also *LAC Minerals, supra* note 3 at 610; *Tlapeck v Chevron Oil Company*, 407 F2d 1129 (1969).

7 JL McDougall, "The Relationship of Confidence" in DWM Waters (ed), *Equity, Fiduciaries and Trusts 1993* (1993) at 165-66.

8 *Id* at 168.

9 [1994] 8 WWR 727 (BCSC) at 738.

The final branch of the test can be met even if there is no evidence of prejudice to the disclosing party. The presence or absence of prejudice is instead treated as a factor which will affect available remedies.[10]

Where a duty of confidence is established, the general rule is that confidential information must be used only for the purposes for which it was imparted until the express or implied consent of the disclosing party is obtained.[11] Implied consent will be deemed to be given, and a receiving party will be released from its common law duty of confidence, where the disclosing party has released the information publicly.[12] Accordingly, the common law duty of confidence may extend for a potentially unlimited duration, which is an extremely unattractive timeframe for receiving parties.

Considerable breadth and flexibility are available to a court in selecting a remedy for breach of the common law duty of confidence, including:

- an injunction to prevent the continued use of the confidential information;[13]
- an accounting of profits;[14]
- a compensatory remedy for the market value of the information misused;[15]
- an award of damages;[16] and
- a constructive trust over any property obtained through the misuse of confidential information.[17]

---

10 *Cadbury Schweppes* [1999] 1 SCR 142*; Pharand, supra* note 5.

11 Saltman, *supra* note 6 at 415.

12 *UK v Observer Ltd* (1988), 99 NR 241 (HL)

13 *LAC Minerals supra* note 3 at 671; *Pre-Cam Exploration & Development Ltd v McTavish* (1966), 57 DLR (2nd) 557 (SCC). See also *Tenatronics Ltd v Hauf,* [1972] 1 OR 329 (Ont HC).

14 *LAC Minerals, id.*

15 *Id; Seager v Copydex Ltd* [1967] 2 All ER 415 (CA).

16 *Cadbury Schweppes supra* note 10; *LAC Minerals, id;* Saltman *supra* note 6.

17 *LAC Minerals, id* at 679. The majority of the Supreme Court of Canada held that a constructive trust was the appropriate remedy. Justice Sopinka and Justice MacIntyre, dissenting in part, argued that damages were the appropriate remedy. Sopinka found no support in the cases for the imposition of a constructive trust over a property acquired as a result of the misuse of confidential information and concluded that constructive trusts were employed only in the context of a pre-existing special relationship (eg, where a fiduciary relationship existed) between the parties. Sopinka stated:

> *A restitutionary remedy is appropriate in cases involving fiduciaries because they are required to disgorge any benefits derived from the breach of trust. In a breach of confidence case, the focus is on the loss to the plaintiff, and as in tort actions, the particular position of the plaintiff must be examined. The object is to restore the plaintiff monetarily to the position he would have been in if no wrong had been committed ... Accordingly, this object is generally achieved by an award for damages* (at 618).

Justice La Forest, writing for the majority of the court, strongly disagreed with this approach, stating:

> *The approach taken by ... Sopinka J., would, in my view, have the effect not of encouraging bargaining in good faith, but of encouraging the contrary. If by breaching an obligation of confidence one party is able to acquire an asset entirely for itself, at a risk of only having to compensate the other for what the other would have received if a formal relationship between them were concluded, the former would be given a strong incentive to breach the obligation and acquire the asset. ... The imposition of a remedy which restores an asset to the party who would have acquired it but for a breach of fiduciary duties or duties of confidence acts as a deterrent to the breach of duty and strengthens the social fabric those duties are imposed to protect* (at 672-673).

Accordingly, the court imposed a constructive trust on the Williams property in favour of International Corona.

This last available remedy, the constructive trust, gives rise to the greatest concern among receiving parties. Their principal worry is that in the event of a dispute, the disclosing party will have a financial incentive to put the receiving party on notice fairly early and then 'lie in the weeds' for as long as possible until just before an action has become barred by statute, while the receiving party expends time, effort and money on the mineral property in question – thereby eliminating the early stage risk associated with the property. If, following the receiving party's expenditures, there appears to be the prospect of a significant mineral discovery on the property, the disclosing party will then, and only then, pursue its legal remedies. In other words, the remedy of a constructive trust effectively gives the disclosing party a 'free option' on a mineral project advanced by the receiving party where there has been a possible misuse of confidential information.

The common law duty of confidence is not a uniquely Canadian concept. As Frank Erisman and John D McCarthy have noted, "American and Canadian law differ little on the theory of breach of confidence."[18] In the United States, the common law duty relating to the protection of confidential information has its basis in the tort of misappropriation of trade secrets.[19] And the three basic elements that must be established to support an action for the misappropriation of trade secrets are largely similar to the three-branch test in *LAC Minerals*.[20] As is the case in Canada, a constructive trust is one of the tools that US judges have at their disposal to remedy a breach of confidence.[21]

Additional protections are afforded in the United States to parties disclosing trade secrets under the Uniform Trade Secrets Act, which has been adopted in 45 states and the District of Columbia.[22] In many respects the Uniform Trade Secrets Act is merely a codification of the existing common law, although it does have some distinguishing characteristics. These principal distinguishing features are that the Uniform Trade Secrets Act:

- does not require a trade secret to have been used continuously to obtain protection;
- requires a plaintiff only to maintain secrecy that is "reasonable under the circumstances"; and
- employs an extremely broad definition of what constitutes a 'trade secret'.[23]

---

18 Erisman and McCarthy, *supra* note 1 at 23-15.
19 Hendricks, *supra* note 1 at 21-15 – 21-16.
20 Restatement of Torts § 757 (1939).
21 A Scott, *Scott on Trusts* § 505 (1967).
22 The states that have not adopted the Uniform Trade Secrets Act are Massachusetts, New Jersey, New York, Texas and Wyoming. For a listing of the states that have adopted the Uniform Trade Secrets Act in one form or another, see Lynn P Hendrix, "Use and Misuse of Confidentiality and Non-disclosure Agreements", *Special Institute on Oil and Gas Agreements: The Exploration Phase* (2004).
23 Hendricks, *supra* note 1 at 21-20.

In addition, several US states have specific criminal statutes which address the theft of trade secrets[24] (a protection which is not available under Canada's Criminal Code).[25]

(b) Fiduciary duty

As broadly as the common law duty of confidence has been extended in cases such as *LAC Minerals*, the obligations on fiduciaries at common law to preserve confidential information are even broader. As the Supreme Court of Canada has noted, "The law takes a hard line against faithless fiduciaries."[26]

Breach of a fiduciary duty may give rise to a cause of action based upon an underlying obligation not to disclose confidential information obtained during the course of the fiduciary relationship. As set out by Justice Wilson in *Frame v Smith*, three criteria must be met in order to establish a fiduciary relationship:

- The fiduciary has scope for the exercise of some discretion or power;
- The fiduciary can unilaterally exercise that power or discretion so as to affect the beneficiary's legal or practical interests; and
- The beneficiary is peculiarly vulnerable to, or at the mercy of, the fiduciary holding the discretion or power.[27]

In *LAC Minerals*, Justice La Forest summarised the principal differences between the common law duties of confidence and fiduciary duties as follows:

- Duties of confidence can arise outside a direct relationship where, for example, a third party has received information from a disclosing party in breach of the disclosing party's relationship to the receiving party, whereas fiduciary duties cannot;
- Breach of confidence will be established only where the innocent party can show that the offending party misused the information, whereas fiduciary law does not require harm to have resulted;
- Remedies from both law and equity are available for breach of confidence, whereas equitable remedies alone are available for breach of fiduciary duty; and
- Although a fiduciary relationship may, and almost always does, co-exist with an obligation to maintain confidence, the reverse is not true.[28]

As noted by Justice Sopinka in *LAC Minerals*, the remedy of a constructive

---

24 For a list of the 23 states that have criminal statutes specifically covering the theft of trade secrets, see Hendricks, *supra* note 22 at 3.

25 *R v Stewart* [1988] 1 SCR 963 (confidential information does not constitute property for the purposes of Section 283 of the Criminal Code (Canada)).

26 *Cadbury Schweppes, supra* note 10 at para 29.

27 [1987] 2 SCR 99 at 136.

28 *LAC Minerals, supra* note 3 at 643-668.

29 *Id* at 623.

trust is available with respect to a breach of fiduciary duties in situations where other remedies would be inappropriate or injustice would result.[29]

Fortunately for most recipients of confidential information in the mining industry, the vulnerability determined to give rise to a fiduciary duty is unlikely to occur in commercial relationships, especially when "dealing with experienced mining promoters who have ready access to geologists, engineers and lawyers".[30] Similarly, in the United States, the fiduciary's burden of proving the utmost faithfulness and fairness to the beneficiary in transactions between them will not be imposed upon parties negotiating at arm's length without a showing of a special relationship between the parties.[31]

While Canadian courts have acknowledged that fiduciary duties are unlikely to arise in most arm's-length commercial mining transactions, courts have also held that a joint venture operator occupies a position of trust as regards its joint venture partners with respect to the management, administration, marketing and sale of products from jointly owned mineral properties.[32] Accordingly, if an operator of a mining joint venture withholds information acquired in its capacity as project manager and uses that information for its own purposes – such as to acquire an adjoining property – a fiduciary obligation has likely been breached. However, fiduciary obligations to maintain confidence do not arise in circumstances governed by a mere option agreement, as opposed to a more fulsome joint venture agreement.[33]

Financial and other professional advisers to mining companies may also have a fiduciary duty of confidentiality, which may be breached if the adviser is in a conflict of interest. In *Gold Reserve Inc v Rusoro Mining Ltd*,[34] the Ontario Superior Court enjoined a hostile takeover bid by Rusoro for the shares of Gold Reserve because Rusoro's financial adviser, Endeavour Financial International Corporation, had a conflict of interest – having formerly acted as Gold Reserve's financial adviser over a four-year period preceding the bid and having accessed Gold Reserve's confidential financial, technical and operational information and prepared confidential cash-flow models for Gold Reserve.

The court found that in advising Rusoro in connection with the takeover bid, Endeavour had breached its obligations under its advisory agreement with Gold Reserve not to disclose confidential information or use it to the detriment of Gold Reserve. Furthermore, even in the absence of the specific prohibitions

---

30 *Id* at 606-607. See also *Cadbury Schweppes supra* note 10 (noting that, in some sense, disclosure of almost any confidential information places the confider in a position of vulnerability to its misuse. However, such vulnerability, if exploited by the receiving party in a commercial context, can generally be remedied by an action for breach of confidence or breach of a contractual term, express or implied. Consequently, a fiduciary duty does not arise from the disclosure of confidential information in and of itself).

31 Erisman and McCarthy, *supra* note 1 at 23-16.

32 See *Visagie v TVX Gold Inc* (1998) 42 BLR (2d) 53 (Ont Ct J (Gen Div)); *Bank of Nova Scotia v Societe Generale (Canada)*, [1988] 4 WWR 232 (Alta CA).

33 *Ontex Resources Ltd v Metalore Resources Ltd* (1993), 13 OR (3rd) 229 (CA), leave to appeal denied, [1993] 4 SCR vi.

34 (2009) 54 BLR (4th) 226 (Ont SC), aff'd 251 OAC (Ont CA).

in the advisory agreement, the court considered that Endeavour's professional advisory relationship gave rise to a fiduciary duty of loyalty and confidentiality: "A person in Endeavour's position must avoid conflicting interests and must not act against the interests of the person confiding in him by utilizing confidential information without the informed consent of that person."[35]

The court also held that absent special measures, such as institutionalised ethical walls, the reasonable presumption is that confidential information will be taken into account and used, whether intentionally or inadvertently, to the disadvantage of the provider of the confidential information.[36]

Accordingly, in control contests or other contentious commercial dealings, mining companies must be careful not merely to avoid misusing confidential information in their own possession, but to ensure that their professional advisers similarly not misuse such confidential information.

(c) The interplay of common law duties and contractual terms

While common law protections against the misuse of confidential information will generally be available in the absence of a confidentiality agreement between parties to negotiations in the mining industry, the converse is not necessarily true. In appropriate circumstances a court may grant relief for breach of confidence on the basis of both breach of contract and one or more common law grounds.[37] The 1998 decision in *TVX Gold* is particularly illustrative of the interplay that can occur between common law principles and the terms of a written confidentiality agreement.

In October 1993 the plaintiffs, which had secured a time-limited exclusive right to negotiate the purchase of the Kassandra mines in Northern Greece from a Greek government-controlled liquidator, provided to TVX, a Canadian public mining company listed on the Toronto Stock Exchange, information about the Kassandra mines and their financial potential under the terms of a confidentiality agreement.

On 25 November 1993 the plaintiffs and TVX entered into a joint venture funding agreement under which they agreed to jointly seek to acquire the mines from the liquidator. TVX was to provide the funding, and the plaintiffs were to receive a 12% carried interest in the mines and an option to take up a further 12% participating interest in the mines once they were acquired.

Once the joint venture funding agreement was in place, TVX assumed control of negotiations with the Greek government and attempted to bid for the mines. However, the Greek government ultimately determined that it was not

---

35 *Id* at para. 58.
36 *Id* at para. 61.
37 See, for example, *Minera Aquiline Argentina SA v IMA Exploration Inc* (2006), 58 BCLR (4th) 217 (BCSC), aff'd, 2007 BCCA 319 (BCCA); *TVX Gold supra* note 32; *Schauenburg Industries Ltd v Borowski Ltd* (1979), 101 DLR (3d) 701 (Ont HCJ).

prepared to allow the mines to be sold privately, but only through a public tender process. As a private transaction could not be reached, TVX terminated its obligations under the joint venture funding agreement with the plaintiffs – as it was explicitly entitled to do under that agreement's terms. TVX then proceeded to bid for the mines on its own in the public tender process. It was ultimately successful, paying $47 million for the mines.

When it was announced that TVX was the successful bidder for the mines, the plaintiffs came forward and claimed their 12% carried interest and the right to purchase a further 12% participating interest. In response, TVX denied that the plaintiffs had any interest in the mines. The plaintiffs subsequently sued for a declaration that TVX had breached its fiduciary duty and common law duty of confidence to them, as well as the terms of their original confidentiality agreement. The plaintiffs also sought a declaration that TVX should hold the Kassandra mines for the plaintiffs' benefit under a constructive trust, subject to reimbursement of TVX's investment to date.

Notwithstanding that the tender process under which TVX acquired the mines was public, an Ontario court held that TVX was not entitled to acquire the mines for its own account, for three principal reasons:

- TVX was in breach of its contractual obligations under the confidentiality agreement when it acquired the mines for itself, and not on behalf of the joint venture, using confidential information that it received from the plaintiffs. Although the information in question was available in the published literature, the information was held to be confidential and have value on the basis of the 'springboard' doctrine, because TVX could not have acquired or produced the information on its own, or at least not as timely or as easily;[38]
- TVX owed a fiduciary duty to the plaintiffs as a result of their relationship as joint venturers. This fiduciary duty required TVX not to pursue the mine except for the joint benefit of both TVX and the plaintiffs, and this duty survived the termination of the joint venture funding agreement;[39] and
- TVX breached its common law duty of confidence owed to the plaintiffs. The plaintiffs had developed credible theories regarding the potential gold reserves on the property and how to exploit them profitably. These theories were confidential to the plaintiffs and those to which they disclosed the information. The information was never made public at any relevant time.[40]

38 *TVX Gold*, *id* at 121-122.
39 *Id* at 122.
40 *Id* at 126.

With respect to the appropriate remedy, the court held that the plaintiffs were entitled to a constructive trust over their original 12% carried interest and, having purported to exercise their option, a further 12% participating interest in the mines upon payment of the costs associated with that interest. Because the fiduciary relationship between the parties developed out of a joint venture where each had the duty to act in the best interests of the joint venture, any acquisition of the target property in breach of such fiduciary duty or in breach of their agreements was an acquisition in trust for the joint venture and owned in accordance with the ownership interests in the joint venture. Therefore, the plaintiffs were not entitled to a remedy of constructive trust being imposed over the entire property.[41]

Decisions such as *TVX Gold* demonstrate how it is necessary that parties understand not only the operation and judicial interpretation of their choice of contractual wording in confidentiality agreements, but also how the common law may affect their circumstances. To reduce the risk of conflict or confusion arising between contractual confidentiality provisions and common law duties pertaining to confidential information, drafters of confidentiality agreements should attempt to ensure as much harmonisation as possible between the contract's terms and the common law.

But what if the parties exchanging confidential information simply do not wish to be governed by the broad protections arising under the common law? The case law suggests that Canadian courts will in most circumstances exclude the availability of common law-based remedies where explicit wording is used in a confidentiality agreement excluding such remedies or duties.[42] In addition, parties can modify a joint venture operator's fiduciary obligations – which give rise to an obligation of confidence – by contract.[43]

3. Principal clauses in confidentiality agreements

3.1 Key functions of confidentiality agreements

As recipients of confidential information in the mining industry will in most cases be wary of subjecting themselves to the expansive scope of the common law duty of confidence, with its broad restrictions on the use of confidential information and extension to even public information under the springboard doctrine, such parties will frequently insist upon the protections of a confidentiality agreement which is specifically tailored to the transaction at hand.

At its core, a confidentiality agreement should perform at least four key functions:

---

41 *Id* at 130.

42 See *Novawest Resources Inc v Anglo American Exploration (Canada) Ltd* 2006 BCSC 769; *Cadbury Schweppes, supra* note 10.

43 See *339765 BC v Tackama Forest Products* (1992), 91 DLR (4th) 129 (BCCA).

- define what information is confidential, and what is not, within the scope of the agreement;
- clearly delineate what the parties can and cannot do with the confidential information;
- specify what happens with the confidential information when the agreement ends; and
- specify what will happen if one of the parties is in breach of the agreement (ie, remedies and enforcement mechanisms).

There may be many other provisions included in a confidentiality agreement, but the foregoing points are likely to be the most important.

(a) Defining confidential information

'Kitchen sink' versus item-by-item approach: Generally, there are two different approaches to defining the scope of information governed by a confidentiality agreement. The parties may apply a 'kitchen sink' approach and define the subject information broadly using a few choice – but very wide-ranging and judicially spring-loaded – terms, such as "all trade secrets and confidential information of the Disclosing Party". Alternatively, parties which are inclined towards more exhaustive definitions may prefer the following wording from the Rocky Mountain Mineral Law Foundation's Form 7:

> *all information, data, knowledge, and know-how (in whatever form and however communicated) relating, directly or indirectly, to Provider (or to is affiliates or to its or their businesses, operations, properties, products, markets, or financial positions) that is delivered or disclosed by Provider or any of its officers, directors, partners, members, employees, agents, affiliates, or shareholders to Recipient in writing, electronically, verbally, or through visual means, or which Recipient learns or obtains aurally, through observation or through analyses, interpretations, compilations, studies, or evaluations of such information, data, knowledge, or know-how.*[44]

The alternative approach to 'kitchen sink' definitions of confidential information is to employ an item-by-item methodology, listing – usually in a schedule – a very narrow range of specific items of information that are to be treated as confidential or a process to be used to specifically identify what material is confidential. Consideration should also be given not only to protecting the information that will be exchanged following execution of the agreement, but also to any information that may have been previously disclosed.

---

44 Rocky Mountain Mineral Law Foundation, Form 7 Confidentiality and Non-disclosure Agreement (1996) at §1.

Although parties may specify in an agreement that certain information is confidential, it does not necessarily follow that the courts will agree. For example, in *Drake International Ltd v Miller*, although the covenant in question referred to "trade secrets, techniques and information" and classified such information as confidential, the court held that this was not determinative, stating that: "All cases agree that matters do not become confidential merely because the contract states them to be so. We must examine the business to determine it for ourselves."[45]

In other words, where the courts reach a determination that a particular item of information is not in fact confidential, as a matter of either law or fact, the mere acknowledgement by one party in a confidentiality agreement that such information should be protected as being confidential is no remedy for such a deficiency.[46]

Another principal risk with the 'kitchen sink' approach to defining confidential information is that an overly broad definition may constitute an unenforceable restraint of trade. The basic premise of the doctrine of restraint of trade is:

> *The public have an interest in every person's carrying on his trade freely: so has the individual. But there are exceptions: restraints of trade and interference with individual liberty of action may be justified by the special circumstances of a particular case. It is a significant justification, and indeed, it is the only justification, if the restriction is reasonable – reasonable, that is, in reference to the interests of the parties concerned and reasonable in reference to the interests of the public, so framed and so guarded as to afford adequate protection to the party in whose favour it is imposed, while at the same time it is in no way injurious to the public.*[47]

Accordingly, parties to confidentiality agreements should make every effort to include in the definition of confidential information only those concepts that have gained judicial approval. If the parties agree among themselves that certain information should be protected under an agreement, but it is unclear as to whether a court would consider the information confidential or unreasonable under the doctrine of restraint of trade, the parties may want to expand or refine the 'use' or 'non-disclosure' provisions of the confidentiality agreement to address these concerns, rather than risk creating an unworkably broad definition of confidential information.

In typical mining industry transactions involving exploration-stage properties, the most common type of confidential information exchanged involves exploration data in the form of maps, charts, logs, reports and other

---

45 (1975), 61 DLR (3d) 420 (Ont HCJ) at 425.
46 *Ebco Industries Ltd v Kaltech Manufacturing Ltd*, [1999] BCJ No 2350 (BCSC) (QL).
47 *Nordenfelt v The Maxim Nordenfelt Guns & Ammunition Co Ltd*, [1894] AC 535 at 565 (HL).

documents containing predominantly geological, geochemical and geophysical data, as well as assay and metallurgical test results. Site visits and visual inspection of drill core and other mineral samples are also common. In such transactions, mid-tier and major producing companies will generally have little appetite to review an exploration company's business strategies, marketing or financial information. Indeed, given decisions such as *TVX Gold*, there is considerable danger to the mid-tier and major producing companies that exposure to the mere 'theories' of the exploration company with respect to future product marketing, engineering, metallurgical processes or business development plans during the course of a due diligence review may foreclose the mid-tier's or major's actual product marketing, engineering, research, metallurgical recovery or business development programmes – in which the mid-tier or major may have invested significant sums – if these programmes should touch upon some of the 'theories' propounded by the disclosing party. Given the relatively limited and straightforward nature of the confidential scientific and technical information that mid-tier and major producing companies will seek in reviewing early stage exploration properties, as well as the difficulties posed by disclosure of marketing, engineering, metallurgical processes or business development plans, it is suggested that the item-by-item approach to defining confidential information is generally most appropriate for such transactions involving early stage mineral properties.

For more advanced and complex transactions – such as transactions involving development or producing properties (where proprietary engineering design or mineral extraction technologies have been developed or are in use), and corporate acquisitions – a broader definition of confidential information may be appropriate.

Following the item-by-item approach to defining confidential information, however, presents its own set of risks. If the disclosing party has not taken adequate care in describing every confidential document that is provided to the receiving party and is genuinely confidential, the disclosing party may unwittingly allow valuable information to become part of the public domain and lose any trade secret protection over such information. Therefore, a disclosing party may want to include a backstop 'other confidential information' clause that would extend the protections under the confidentiality agreement to information within a limited category (eg, exploration data) that is transferred to the receiving party, but not specifically identified within the confidentiality agreement, provided that the disclosing party identifies such information as confidential to the receiving party at the time of delivery.

Definitions articulated by the courts: In applying a 'kitchen sink' approach to the definition of 'confidential information', confidentiality agreements will often rely on broad generic terms. Parties drafting such confidentiality agreements, however, should be alert to how courts have interpreted many of these phrases and recognise that certain terms may have different meanings (and may or may not be enforceable) depending on which side of the 49th parallel one sits.

While the term 'confidential information' is not capable of a precise definition, in *Pharand Ski Corp v Alberta*, the court set out a lengthy series of non-exhaustive factors to consider when determining whether information has the necessary quality of confidentiality:

> *A list of factors to be considered to determine if the information has a quality of confidence about it may be found in Ansell Rubber Co. Pty Ltd., v Allied Rubber Industries Pty Ltd., [1967] V.R. 37, and Deta Nominees Pty Ltd v Viscount Plastics Products Pty Ltd., [1979] V.R. 167 at p. 193 … They are:*
>
> *(1) the extent to which the information is known outside the owner's business*
>
> *(2) the extent to which it is known by employees and others involved in the owner's business*
>
> *(3) the extent of measures taken by him to guard the secrecy of the information*
>
> *(4) the value of the information to him and his competitors*
>
> *(5) the amount of money or effort expended by him in developing the information*
>
> *(6) the ease or difficulty with which the information could be properly acquired or duplicated by others [i.e. by their independent endeavours].*
>
> *These factors are useful, but as Fullager, J. states in Deta Nominees [1979] V.R. 167 at p.193 they are not exhaustive:*
>
>> *"I apprehend that neither he [namely Gowans, J. in Ansell's case] nor McInerney, J. in Mense v Milenkovic [1973] V.R. 784 at 796-8 intended to convey that one should slavishly check off the factors against the information, as if one were counting spots on some strange creature to see if it was indeed the species of leopard illustrated in the picture book.*[48]

'Confidential information' is often defined to include trade secrets. The following definition of 'trade secrets', accepted by Justice Chevrier in *RL Crane Limited v Ashton*, provides a useful guideline as to what constitutes a trade secret in Canada (and generally mirrors US law):[49]

> *1st A trade secret … is a property right, and differs from a patent in that as soon as the secret is discovered, either by an examination of the product or any other honest way, the discoverer has the full right of using it … Progress Laundry Co. v Hamilton, 270 S.W. 834, 835, 208 Ky. 348.*
>
> *2nd A trade secret is a plan or process, tool mechanism or compound known only to its owner and those of his employees to whom it is necessary to confide it.*

48 *Pharand, supra* note 5 at 316.

49 For a comprehensive treatment of the law of trade secrets in the United States, see Roger M Milgram, *Milgram on Trade Secrets*(2004); Melvin F Jager, *Trade Secrets Law* (2003); and Henry F Perrit, Jr, *Trade Secrets – A Practitioner's Guide* (1994 & Cum Supp). For a useful analysis of US trade secret law on a state-by-state basis, see Brian M Malsberger, *Trade Secrets: A State-by-State* (2003).

Cameron Mach Co. v Samuel M. Longdon Co, N.J. 115 A. 212, 214; Victor Chemical Works v Iliff, 132 N.E. 806, 811, 299 Ill.532.

3rd The term 'trade secret', as usually understood, means a secret formula or process not patented, but known only to certain individuals using it in compounding some article of trade having a commercial value, and does not denote the mere privacy with which an ordinary commercial business is carried on. Glucol Mfg. Co. v Shulist, 214 N.W. 152, 153, 239 Mich. 70.

4th A trade secret may consist of any formula, pattern, device, or compilation of information which is used in one's business, and which gives him an opportunity to obtain an advantage over competitors who do not know or use it. A trade secret is a process or device for continuous use in the operation of the business. The subject matter of a trade secret must be secret. Restatement, Torts, 757.[50]

Within the resource industries, geophysical and geological data obtained at the owner's expense and effort will likely be found to constitute a trade secret entitled to protection as confidential information.[51]

Canadian law distinguishes between 'trade secrets' and 'know-how' – the latter being information gained by experience or skill. The distinction often arises in the context of post-employment competition where a former employer is seeking to restrain an employee from using or divulging confidential information in competitive employment. Anglo-Canadian jurisprudence has for the most part held that general know-how is not capable of restrictive contractual treatment in the same way as truly confidential information can be.[52] This contrasts somewhat with the approach in the United States, where know-how can constitute a trade secret provided that it:

- is not generally known and not readily ascertainable;
- has independent economic value; and
- is subject to reasonable efforts to maintain secrecy.[53]

Accordingly, US counsel accustomed to incorporating know-how within the definition of confidential information[54] should be careful to avoid the use of such terms in agreements governed by Canadian law – or, at a minimum, restrict contractual protections to 'specialised' know-how.[55]

US confidentiality agreements frequently extend contractual protections to 'proprietary information' and US authorities have confirmed that such

50 *RL Crane Limited v Ashton*, [1949] OR 303 at 388-89.

51 See *Tlapek v Chevron Oil Co*, 407 F 2d 1129 (8th Cir 1969); see also *Hunter v Shell Oil Co*, 198 F2d 485 (5th Cir 1952); *Ohio Oil Co v Sharp*, 135 F 2d 485 (10th Cir 1943).

52 See *Faccenda Chicken Ltd v Fowler* [1986] 1 All ER 617 (CA); *Pizza Pizza Ltd v Gillespie* (1990), 75 OR (2d) 225 (Ont Ct (Gen Div)).

53 Hendricks *supra* note 1 at 2.

54 For example, Form 7, *supra* note 44 at §1 (incorporating know-how within the definition of confidential information).

55 See John T Ramsey, "Drafting Confidentiality Agreements in Canada" (1989) 4 *IPJ* 157 at 165 (drawing a distinction between 'specialised' know-how, which may be deserving of protection, and 'general' know-how, which is not).

protections can be enforceable, even where the proprietary information does not rise to the level of a trade secret.[56] By contrast, in Canada, there does not seem to be 'property' in confidential information.[57] Such information has value and is protected only to the extent that it remains secret and to the extent that the courts are willing to enforce obligations, express or implied, of confidence.[58] Accordingly, parties should be wary of relying on the concept of proprietary information in confidentiality agreements governed by Canadian law.

Often, confidentiality agreements will restrict not only the disclosure of confidential information, but also disclosure of the very existence of the confidentiality agreement or any contemplated transactions between the parties. The Form 7 wording on this point is as follows: "Neither Recipient nor Provider shall, directly or indirectly, disclose to any third party the terms and conditions of this Agreement or the transactions that are the subject of this Agreement, without the other party's prior written consent."[59]

Disclosing parties will generally insist upon such covenants to ensure some degree of control over both the due diligence process and any communication to the markets concerning such process. A 'transactional information' covenant will also protect against possible collusion among potential bidders for a property, asset or company. Receiving parties, on the other hand, will want to ensure that any restrictions on the disclosure of such 'transactional information' is a mutual obligation between the parties, as it reduces the risk that the recipient of information will be used as a stalking horse to flush out a competing transaction.

Disclosure medium – written versus oral or visual: While tracking and recording the disclosure of written information and data is a generally manageable task, the provision of confidential information orally or visually (eg, visual inspection of drill core) can present potential difficulties. Recipients of information frequently object to the protection of oral and visual data because of the difficulty in proving whether a particular item of information was in fact disclosed, should a dispute arise in the future. Recipients, however, should also recognise that excluding oral or visual information from coverage under a confidentiality agreement may reduce the ease and efficiency of the due diligence process if the disclosing party is unable to respond to telephone inquiries, make management presentations or participate in interviews or Q&A sessions. One solution occasionally employed to address these difficulties is to require the disclosing party to confirm any confidential information that is

---

56 Ferguson, *supra* note 1 at 7-6.
57 See RJ Roberts, "Is Information Property?" (1987), 3 *IPJ* 209; Marie Bourgeois, "Protecting Business Confidences: A Comparative Study of Quebec and French Law" (1987), 3 *IPJ* 259.
58 Ramsay, *supra* note 55 at 164.
59 *Supra* note 44 at §8.

orally or visually disclosed in writing within a reasonably short period after its disclosure. Both disclosing and receiving parties can benefit from the creation of such a written record, as this process will reduce the likelihood of any future dispute turning upon the memories of personnel involved in the transaction or their credibility on the witness stand.

Disclosure mechanics: Most confidentiality agreements will specify in some detail the manner in which confidential information is to be communicated between the parties. Important considerations include whether the confidential information is to be provided all at once or whether it will be provided in a piecemeal fashion. Will the confidential information be made available in a single location – such as an electronic data room – or multiple locations, or will it be physically delivered to the recipient? Over what timeframe will information be made available? How is the recipient to take delivery or receive the confidential information? What physical form will the confidential information take (eg, electronically stored, documentary, oral presentation)? Is the recipient permitted to copy the information?

Parties should also consider whether their confidentiality agreement should specify any means by which the information will be identified as being confidential. Apart from the practical considerations in managing the data, there is an important legal reason in Canada for disclosing parties to identify their information as confidential. It has been held that confidential information must be distinguishable from the range of information which is generally available in the industry if a confider hopes to receive the assistance of the courts in protecting it.[60] Commonly employed identification mechanisms include:

- preparing and updating a 'master list' or 'control sheet', which identifies each item of confidential information disclosed, copies made and individuals who had access to the information – which has become logistically easier in the era of electronic data rooms;
- requiring confidential documents to be stamped or watermarked with a cautionary legend, such as 'confidential' or 'do not copy';
- numbering copies of confidential documents; and
- restricting and monitoring physical or electronic access to the confidential information.

Disclosing parties, however, should be wary of hardwiring unnecessarily complicated disclosure mechanics into their confidentiality agreements. If not complied with in practice, such clauses may unintentionally result in a loss of

60 *Faccenda Chicken, supra* note 52 at 627; *Terrapin Ltd v Builders Supply Co (Hayes) Ltd,* [1967] RPC 375 at 391; *Printers and Finishers Ltd v Holloway*, [1964] 3 All ER 731 at 738 n.

protection. In *United Technologies Corp v Platform Computing Corp*, for example, an Ontario court ruled that failure to follow identification mechanisms explicitly set out in a confidentiality agreement – in this case, marking documents as 'confidential' – was evidence that the parties intended the unmarked data to be provided without any restrictions as to use.[61]

Derivative works: As geology is an interpretative science, mining industry participants may wish to consider whether their confidentiality agreements should extend to 'derivative works'. Such works may include interpretative reports, assay checks, engineering, metallurgy and environmental studies, marketing studies, economic and sensitivity analyses, financial models and other data generated by the receiving party or the receiving party's consultants and contractors, based in whole or in part on the raw data provided or site visits to a property. If the provisions of a confidentiality agreement extend to 'derivative works', the parties should carefully consider whether the nature of such derivative works warrants the same treatment as other confidential information at the termination of the due diligence process. If derivative works are to be returned to the original disclosing party, should the party that prepared the derivative works be compensated for undertaking and sharing its analysis? What if the derivative work involves the application of engineering or metallurgical processes, marketing knowledge or regional geographical information that is itself confidential to the party generating the derivative works? In such circumstances, destruction, rather than a return of the derivative works, may be warranted to preserve the reasonable expectations of confidentiality of both parties.

(b) Exclusions from the definition of confidential information

In addition to specifying what categories or items of information are subject to protection, a confidentiality agreement should identify what information is excluded from such protection. Standard exclusions in the mining industry focus on information that:

- is already in the possession of the receiving party at the time the information was disclosed;
- is already in the public domain at the time the information was disclosed or subsequently becomes part of the public domain through no violation of the confidentiality agreement;
- is subsequently disclosed to the receiving party from a third party which did not acquire such information from the disclosing party under a contractual obligation of confidence; or
- is developed independently by the receiving party without the use of any confidential information provided by the disclosing party.

61 [1998] OJ No 883 (Ont Ct (Gen Div) (QL), rev'd on other grounds [1999] OJ No 4490 (Ont CA) (QL).

Information already known to the receiving party: Although information disclosed under a confidentiality agreement may not be in the public domain, the receiving party may have developed the same information on its own prior to disclosure by the disclosing party. This clause protects against the receiving party losing its control over such information in those circumstances.

Notwithstanding the presence of such a clause, it would be prudent for a receiving party to confirm as soon as practicable in writing to the disclosing party the existence of any overlapping confidential information that the receiving party possessed prior to any disclosure of information under the confidentiality agreement, given that the onus will likely fall on the receiving party to prove that the information falls within this exclusion in the event of a dispute.

Some disclosing parties will insist on modifying this exclusion with the addition of wording to the effect that the information "was not, directly or indirectly, acquired under an obligation of confidentiality to the disclosing party", to ensure that any unknown prior breach of a contractual or common law duty of confidence between the parties is not effectively waived.

Public domain: A receiving party can contractually bind itself under a confidentiality agreement not to use or disclose information in the public domain,[62] although in practice courts will not generally restrain the use of public information except where the disclosing party has a very clear moral claim to bind the conscience of the receiving party.[63] To avoid the possibility of a receiving party becoming restricted from using publicly available data, information already in the public domain will invariably be specifically excluded from the definition of protected information under confidentiality agreements. However, disclosing parties will likely insist that information not be treated as falling within the public domain under this exclusion if it has become public through some fault of the receiving party, to avoid rewarding such a party for breach of the disclosure restrictions under a confidentiality agreement.

But when does information enter the public domain? In the absence of explicit wording within a confidentiality agreement, the jurisprudence on this question is mixed. In the *Ridgewood Resources* decision, for example, the Alberta Court of Appeal held that information had become public where it had become known to just over 20 people.[64] By contrast, the Delaware Circuit Court held in *Data General Corp v Digital Computer Controls*, Inc that a distribution of between

62 *Ridgewood Resources supra* note 5 at 503 ("The communication of information which is in the public domain, though not known to the recipient, does not (*apart from contract*) make the recipient liable for its use" [emphasis added]).

63 *Schering Chemicals Ltd v Falkman Ltd*, [1981] 2 All ER 321 at p338, [1982] QB 1 (CA).

64 *Ridgewood Resources, supra* note 5 at 503.

80 and 6,000 copies of proprietary drawings did not terminate their relative secrecy and non-public character.[65]

Adding to this uncertainty is the question of which persons constitute the public for the purposes of determining confidentiality. Under Canadian law, the answer is not necessarily intuitive. The jurisprudence establishes that it is the 'relevant public', not merely the 'general public', that must be considered. In *TVX Gold,* for example, Justice Feldmen held: "when examining the confidentiality of information in the context of its notoriety or availability in the public domain, one looks at the relevant public as opposed to the general public which has no interest in or reason to have access to such information."[66]

In *TVX Gold,* the court found that the 'relevant public' comprised interested mining companies and, as some of the documents claimed by the disclosing party to be confidential were generally available to these other mining companies, such documents were not protected by the confidentiality agreement entered into between the parties to the dispute.

In addition, in many circumstances information disclosed to a receiving party will be mixed in character: some of the information will fall within the public domain and some will not. In such circumstances Lord Denning identified the receiving party's obligations to be as follows: "When the information is mixed ... the recipient must take special care to use only the material which is in the public domain. He should go to the public source and get it: or at any rate, not be in a better position than if he had gone to the public source. He should not get a start over others by using the information which he received in confidence."[67]

If the term of a confidentiality agreement is expected to continue for an extended length of time, receiving parties will also want to exclude from the definition of confidential information those documents and data that become public after commencement of the agreement, so as not to be placed at a competitive disadvantage with respect to the use of such information.

Information subsequently disclosed by a third party: *Obiter dicta* in the *Pharand* decision suggest that a receiving party may not be released from its disclosure restrictions under a confidentiality agreement where information is provided to the disclosing party – or even enters the public domain – as a result of disclosure by a third party.[68] In *Pharand,* Justice Mason quoted from the House of Lords decision in *UK v Observer Ltd,* which held:

Information may lose its original confidential character if it subsequently enters

---

65 188 USPQ 276 (Del Cir Ct, 1975) at 279.

66 *TVX Gold, supra* note 32 at 115. See also *Mid-Canada Millwork Ltd v Delano Building Products Ltd,* (1997) 119 Man R (2d) 67 (QB), *aff'd,* (1998), 126 Man R (2d) 311 (CA), leave to appeal dismissed, [1998] SCCA No 399.

67 Seager, *supra* note 15.

68 *Supra* note 5 at 320.

the public domain. If the confider publishes the information this releases the confidant from his duty of confidence ... the courts have, however, so far refused to extend this principle where the confidential information is published by a third party ... or to the case of publication of the information by a confidant.[69]

Accordingly, receiving parties will generally insist upon excluding from the definition of protected information any data which is subsequently disclosed to the receiving party from a third party. This will prevent the receiving party from being placed at a competitive disadvantage by being restricted from entering into transactions with a third party which has developed similar information independently of the disclosing party.

Receiving parties should avoid limiting these exclusion clauses to information obtained by a third party "under an obligation of confidence" to the disclosing party. This is because the receiving party may be unable to conclusively determine whether information supplied to it by a third party is subject to a fiduciary or other vague common law obligation of non-disclosure. Instead, a more reasonable and more easily ascertainable limitation to this exclusion would restrict the excluded information to data that was not acquired by the third party under a 'contractual obligation of confidence'.

Information developed independently: The most contentious category of frequently excluded information involves information that is developed independently by the receiving party without the use of any confidential information provided by the disclosing party. A claim that a receiving party's interest in a recently staked nearby mineral property conveniently arose from other independently developed sources, while the receiving party was also in possession of confidential information provided by the disclosing party that indicated the prospective nature of the property, will often strain credulity unless a well-defined internal 'firewall' procedure can be documented by the receiving party. This likely explains the absence of this exclusion from Form 7.[70] To address these concerns, the parties may wish to limit an 'independent development' exclusion to those subject areas in which the receiving party specifically identifies that it has an active development programme ongoing (eg, a proprietary mineral extraction technology development programme or a regional exploration programme).

Burden of proof: The burden of proof with respect to whether information is or is not subject to one of the exclusions from the definition of confidential information should be specifically addressed in the confidentiality agreement. Generally, the disclosing party will require the receiving party to bear this

---

69 *Supra* note 12 at 254-5.
70 *Supra* note 44 at §1.

burden of proof, as the disclosing party will likely have an insurmountable difficulty in establishing what was or was not within the prior knowledge or independently developed by the receiving party.

(c) Restrictions on disclosure

Traditional variants on protective measures: The degree of care required to be exercised and the particular protective measures required to be instituted by a receiving party with respect to confidential information typically vary from confidentiality agreement to agreement. Five traditional contractual variants of the standard of care are generally applied in maintaining confidentiality. Listed in order of increasing severity, these standards are as follows:

- Hold the confidential information in confidence in accordance with industry standards: The difficulty with this standard of care is that there is little or no readily available evidence as to the most common procedures in place across the mining industry to protect confidential information. Until such time as standard confidentiality procedures are adopted across the industry,[71] this standard of care will be subject to considerable uncertainty and in a dispute will likely turn on a court's assessment of conflicting expert testimony as to what standards of treatment of confidential information are most common in the industry.
- Hold the confidential information in confidence, exercising the same degree of care that the recipient uses for its own confidential information: If the disclosing party is confident that the receiving party maintains a high level of internal security over its confidential information, it may find this standard of care to be acceptable. This standard has the logistical advantage of permitting the receiving party to protect the confidential information received from the disclosing party through existing systems without having to adopt any additional precautions. The disclosing party should consider requesting and reviewing any written procedures or policies of the receiving party with respect to its internal treatment of confidential information prior to accepting this standard, or obtaining representations in the confidentiality agreement as to the policies and procedures of the disclosing party for maintaining the confidentiality of information.
- Take reasonable precautions, reasonable efforts or best efforts to hold the confidential information in confidence: This language establishes a broad, objective standard with which courts and arbitrators will be familiar – the efforts that a prudent person would take, acting reasonably, to hold the disclosed information in confidence. Sometimes

71 For example, ISO 17799: Information Security Management (2005).

this standard is further clarified in confidentiality agreements with a recital or schedule setting out a number of specific 'reasonable steps' to be undertaken.[72] Receiving parties, however, should be wary of formulations of this standard which require the application of 'best efforts' – as that standard of conduct is considerably more stringent that a standard of reasonableness.[73]

- Hold the confidential information in confidence or keep it 'strictly confidential': Receiving parties should be cautious of this wording if it appears without further qualifications. This standard can be read to establish strict liability, regardless of fault or negligence, and oblige the receiving party to effectively guarantee that there will be no disclosure under any circumstances whatsoever. This is more than any receiving party, or even the disclosing party itself, can reasonably do, as there always remains the possibility that a dishonest employee or an outsider may circumvent even the most stringent security systems and procedures and gain access to the information. This standard of care will generally be appropriate only if the obligation of confidentiality is further defined or modified or if the receiving party is exculpated from liability for disclosures beyond its reasonable control.
- Hold the confidential information, or the benefits from its improper use, in trust for the disclosing party: Receiving parties should reject this wording, as it will raise their obligations to the disclosing party to the standard of a fiduciary under which even slight or innocent lapses in confidentiality will expose the receiving party to liability.

---

72 An example of such a recital or schedule is as follows:

Without restricting the generality of the foregoing, the reasonable steps that the Recipient shall take include:

(a) physical security of areas where access may be gained to Confidential Information;

(b) security measures for electronic storage and transmission of data including or derived from Confidential Information;

(c) controls on access of any computer facility and tape or disk library where any Confidential Information may be kept;

(d) visitor control;

(e) controls over photocopying Confidential Information;

(f) document and computer network control systems which limit access to the Confidential Information to employees and agents who have a need to have such access and which provides for a secured method of destruction of sensitive data;

(g) confidential agreements with the Recipient's employees, agents or invitees who are permitted access to Confidential Information, which agreements shall be

(i) for the benefit of both the Recipient and the Disclosing Party;

(ii) enforceable, at the option of the Disclosing Party;

(A) by the Recipient at the expenses of the Recipient, or

(B) by the Disclosing Party

and in either case the Recipient shall cooperate fully with the Disclosing Party in enforcing the terms of such agreements.

Ramsay, *supra* note 55 at 176.

73 See, for example, *Atmospheric Diving Systems Inc v International Hard Suits Inc* (1994), 89 BCLR (2d) 356 (BCSC) ('best efforts' include doing everything known to be usual, necessary and proper for ensuring the success of the endeavour and means taking, in good faith, all reasonable steps to achieve the objective, carrying the process to its logical conclusion and "leaving no stone unturned").

To further define the receiving party's obligations and liabilities, confidentiality agreements may provide that if the specific protective measures mandated by the agreement are in place, the receiving party shall not be held responsible if any confidential material should be divulged to a third party, whether by reason of honest mistake, dishonest appropriation or otherwise.

'Need to know' disclosure to representatives: Careful thought should be given to which representatives of the receiving party will be entitled to access information under the terms of a confidentiality agreement. To preserve flexibility, confidentiality agreements generally provide that information will be disclosed only to those representatives of the receiving party who have a *bona fide* 'need to know' the information in order for the receiving party to complete its due diligence and proceed with a transaction. As the mining industry involves a close interdependence on, and requires the participation of, parties from many specialised disciplines, it will often be necessary for confidential information to be shared beyond the receiving party's own directors, officers and employees. Accordingly, it is often prudent to expand the categories of representatives to which confidential information may be disclosed to include third-party contractors, consultants, professional advisers, banks, underwriters and their respective professional advisers.

Concerns over privity and enforcement of confidentiality obligations with regard to a receiving party's representatives are typically addressed by requiring that a receiving party's representatives which are provided with confidential information be informed of the confidential nature of the information and the terms of the original confidentiality agreement. Generally, as a matter of Canadian law, such procedures will impose a common law obligation of confidence upon the third-party recipient, which can be pursued before the courts.[74] It also may be advisable to evidence that the third party has been so informed through a written undertaking or agreement.

Regardless of whether second-tier acknowledgements or undertakings of confidentiality are obtained, a disclosing party should require that the receiving party assume responsibility for, and indemnify the disclosing party against, any breach by its representatives or former representatives.

Disclosure required by law, court order or stock exchange requirements: Confidentiality agreements should always provide that the receiving party may disclose confidential information required by law, court order or any stock exchange having jurisdiction. Disclosing parties, however, will sometimes insist

74 See, for example, *Cadbury Schweppes*, *supra* note 10 at 157 ("Equity will pursue the information into the hands of a third party who receives it with the knowledge that it was communicated in breach of confidence (or afterwards acquires notice of that fact even if innocent at the time of acquisition) and impose its remedies").

that a written opinion to such effect be obtained from the receiving party's counsel as a condition precedent to any such disclosure. Consideration should also be given as to whether the disclosing party should be provided with a contractual right to:

- become a party to, or defend, any action compelling disclosure of the information; or
- be promptly advised of the disclosure requirement prior to disclosure being made, to permit the disclosing party to seek a protective order concerning the information.

The parties may also wish to agree to cooperate in good faith, and to keep each other informed of developments concerning and use reasonable commercial efforts to limit the extent of disclosure of confidential information required by law, court order or any stock exchange having jurisdiction. However, covenants requiring advance notice to disclosing parties or requiring receiving parties to cooperate with disclosing parties to limit disclosure should not be applicable in circumstances where a representative of the receiving party may be compelled to make disclosure or stand liable for contempt or suffer some other form of censure or significant penalty.

Where the required disclosure involves a press release mandated under securities laws or stock exchange requirements, confidentiality agreements will usually impose an obligation to provide the other party with prior notice of the intent to issue, and an opportunity to review a draft of, any press release. There will often be a corresponding right for the other party to comment on the release within a very limited timeframe. However, public companies must be careful that their confidentiality agreements do not leave decisions concerning press releases to mutual agreement – as a failure to reach mutual agreement may place a party in the untenable position of having to choose between breaching its obligations under securities laws and stock exchange requirements or breaching its obligations under the confidentiality agreement.

(d) Representations and disclaimers as to confidential information

Disclosing parties typically seek, in their confidentiality agreements, to broadly disclaim any representations and warranties, express or implied, as to the accuracy, completeness or usefulness of the confidential information that is subject to the agreement, on the basis that the scope of any such representations and warranties is best negotiated at the time the parties are negotiating a definitive agreement governing a transaction – which will generally occur after due diligence is concluded.

In many confidentiality agreements, however, the disclosing party's disclaimers are so expansively worded that they not only disclaim warranties, but disclaim all liability, even for knowing misstatements. It is submitted that such overly broad exclusions of liability are inappropriate, as these disclaimers are intended to protect (temporarily, pending a definitive agreement) the disclosing party from liability for inadvertent errors or omissions in disclosure at the early stages of a transaction.

Indeed, there are a limited number of representations and warranties that a receiving party can, and arguably should, legitimately seek from a disclosing party in a confidentiality agreement which do not offend the underlying basis for the usual disclaimer of representations and warranties as to the accuracy, completeness or usefulness of information. These representations and warranties include the following:

- The confidential information disclosed does not contain any knowing misstatements;
- The disclosing party has not withheld any information which materially qualifies or contradicts the confidential information disclosed to the receiving party. Where a receiving party expends its time, effort and opportunity costs in reviewing confidential information in a business transaction, it is a reasonable expectation that it will be provided with balanced data, reflecting both the strengths and the weaknesses of a project; and
- The disclosing party has the right to disclose the confidential information to the receiving party. A receiving party will not want to receive information if such disclosure may breach a duty of confidence owed by the disclosing party to a third party (eg, an optionor, royalty holder or a joint venture partner) and expose the receiving party to claims that it owes broad common law duties of confidence to the third party in respect of the information. Such a representation and warranty, however, will have value only if supported by a meaningful indemnity. If the disclosing party does not have the financial resources to satisfy an indemnity in the event that a suit is brought against the receiving party by the 'partner' of the disclosing party alleging a breach of common law duties of confidence, the receiving party may be better off forgoing such a representation and warranty and seeking a direct agreement with such 'partner' concerning the confidential information (perhaps by making the 'partner' a party to the confidentiality agreement with the disclosing party).

Finally, confidentiality agreements should provide that any disclaimer of representations and warranties by the disclosing party is subject to the terms of any definitive agreement ultimately entered into by the parties.

(e) Uses of confidential information

Restricted uses: Often the most contentious clause in any confidentiality agreement is the 'restricted use' clause – which provides that confidential information will be used solely for a particular purpose, such as to evaluate a possible transaction, and that all other uses are prohibited.

Disclosing parties have traditionally insisted upon 'restricted use' clauses for two reasons. First, under US law, granting broad or unlimited rights of use may result in a loss of trade secret protection under applicable trade secret laws.[75] Second, disclosing parties will not want a receiving party to somehow apply the disclosing party's confidential information for some unexpected purpose that results in either a detriment to the disclosing party or an unjust enrichment of the receiving party.

The British Columbia Supreme Court's decision in *Aquiline* is the most noteworthy decision on restricted use clauses in the mining industry. The dispute between the parties in *Aquiline* received considerable press attention prior to trial[76] and the initial trial court decision was described as both a "far reaching"[77] judgment and one that "might make members of the mining industry run to review their outstanding agreements".[78]

Aquiline involved the sale by Newmont Gold Corporation of an Argentine subsidiary, Minera Normandy Argentina SA, which was the owner of a series of mineral claims – known as the Calcatreu Project – which covered approximately 730 square kilometres in Rio Negro province and upon which a modest gold resource had been discovered.

Newmont arranged for various prospective purchasers – including the defendant, IMA Exploration Inc – to receive an information brochure describing the Calcatreu Project and, if desired, participate in site visits, provided that such parties signed a confidentiality agreement with Newmont and Minera.

On 6 September 2002 IMA, Newmont and Minera entered into a confidentiality agreement. The confidentiality agreement was governed by Colorado law and applied to all confidential information "concerning" and communications "relating to" the Calcatreu Project. The agreement contained several vague restricted use clauses – noting that confidential information was being provided "in connection with Reviewer's review of the Project" and that "Confidential Information provided by Newmont to Reviewer will be used by Reviewer or Reviewer's Representatives only for the purpose of the Project". The confidentiality agreement also contained an "Area of Exclusion" clause, which

---

75 Hendrix, *supra* note 1 at 9.

76 See, for example, Eric Reguly, "Investors roll dice in silver feud", *Globe and Mail* (13 October 2005) B2; Eric Reguly, "Former Tailor Trying to Sew up Bonanza", *Globe and Mail* (14 June 2005) B2.

77 Carmen L Diges, "Recent Developments in Mining", *Lexpert 500* (2007).

78 Josh Lewis, "Confidentiality Agreements – Part 2", *CIM Magazine*, Feb 2007 at 50.

prohibited recipients of confidential information from acquiring, directly or indirectly, any mining claims, permits, concessions or other property situated within two kilometres of and parallel to all exterior boundaries of the Calcatreu Project.

During a September 2002 site visit, a geologist with IMA requested additional information respecting the Calcatreu Project, including approximately 500 stream sediment samples within and around the Calcatreu Project (the 'BLEG B data'), which was provided to him. During this site visit, the geologist also happened to notice a map in Minera's offices depicting the location of the samples used to generate both the BLEG B data and approximately 1,000 other stream sediment samples over an area of approximately 12,000 square kilometres which extended up to 100 kilometres outside of the southern boundary of the Calcatreu Project (the 'BLEG A data'). After a discussion concerning the map, IMA's geologist requested the BLEG A data which was provided during a second site visit in late October 2002. No other potential purchaser was provided with the BLEG A data. On 6 November 2002, before any IMA geologists could review the BLEG A data, IMA advised Newmont that it would not be bidding on the Calcatreu Project.

In early November 2002 Aquiline Resources Inc offered to purchase the Calcatreu Project from Newmont for US$2 million. At the time of the offer, Aquiline had not conducted a site visit or any due diligence, other than a review of the information brochure. Following discussions with Newmont, Aquiline agreed to restructure its offer for the Calcatreu Project as a purchase of Minera and increase its bid by US$50,000. By structuring the transaction as a corporate purchase, Aquiline obtained effective beneficial ownership of the BLEG A data and the benefit of Minera's confidentiality agreement with IMA. On 28 January 2003 Aquiline entered into a purchase agreement with Newmont and the transaction closed on 10 July 2003.

In late November 2002 IMA's geologist – who still retained the BLEG A data, but had not yet reviewed it – asked IMA's management if IMA's arrangements with Newmont permitted him to look at the BLEG A data. Not receiving a response, he reviewed the BLEG A data and noticed a cluster of silver anomalies in North-central Chubut province, approximately 40 kilometres outside the boundaries of the Calcatreu Project and 38 kilometres outside the area of exclusion. On 6 December 2002 IMA staked the area around the silver anomalies, which became known as the Navidad Project. On 3 February 2003 IMA announced that it had found a "Bonanza Grade Silver – Copper – Lead Discovery" on the Navidad Project. Subsequent drilling and exploration of the property would delineate an indicated resource of 305 million ounces of silver and 2.9 billion pounds of lead.

In May 2003 Aquiline examined the BLEG A data, discovered the silver anomalies and expressed its own interest in staking the area – only to discover that the property had already been staked by IMA, the only other bidder to have seen the BLEG A data. In March 2004 Minera commenced legal proceedings against IMA in the British Columbia Supreme Court.

The court gave an expansive interpretation as to the restricted use clause, based on a number of precedents,[79] on the basis that terms used in the confidentiality agreement such as "in connection with . . . the Project", "concerning the Project" and "relating to the Project" – although seemingly innocuous connecting phrases on their face – were extremely broad in scope as a matter of contractual interpretation and were to be construed as applying to any information that was provided in the context of IMA's evaluation of a possible transaction concerning the project.[80]

Based also on its broad interpretation of the terms 'confidential information' and 'project', the court concluded that the use by IMA of the BLEG A data to find and stake the Navidad Project was in clear breach of the restricted use provisions of the confidentiality agreement – as IMA had used the BLEG A data for its own staking purposes and not for the purpose of the Calcatreu Project.[81] Justice Koenigsberg also made it clear that such a restricted use provision did not represent such an unworkably broad restriction on parties which enter into confidentiality agreements that it would harm the mineral exploration business generally. The court reasoned that the exclusions to the definition of confidential information – relating to independently developed or non-confidential information – would permit a receiving party to stake prospective ground in any region, even a region for which anomalous sample data was disclosed by the disclosing party, provided that such staking was based on the receiving party's own data or public information and not on the disclosed data.[82]

As for the two-kilometre area of exclusion provision in the confidentiality agreement, the court critically held – again relying in part upon the scholarly literature[83] and in part upon the fact that the area of exclusion provisions made no explicit reference to 'confidential information' – that the area of exclusion clause was an independent, standalone covenant, which in no way limited or coloured the interpretation of the restricted use provisions.[84]

In other words, the *Aquiline* decision essentially stands for the proposition that a broadly worded restricted use clause effectively serves as an area of exclusion clause of potentially unlimited geographic scope in circumstances

---

79 *Supra* note 37, 58 BCLR (4th) at 237, citing *Nowegijick v R*, [1983] 1 SCR 29; *Slattery (Trustee of) v Slattery*, [1993] 3 SCR 430.

80 *Id.*

81 *Id* at 244.

82 *Id* at 241-242.

83 *Id* at 240-241, citing M Hardwicke-Brown, "Confidentiality and Dispositions in the Oil and Gas Industry" (1997) 35 *hh* 356 at 387.

84 *Id* at 240-242.

where a property is staked in reliance upon misused confidential information. Arguably, a recipient of technical data could stake ground 2,000 kilometres and a continent away from a property which is the subject of a confidentiality agreement; but if that staking was in any way, shape or form based on confidential information, then it would constitute a breach of the agreement. This represents a significant development in, and broadening of, the interpretation of confidentiality agreements in Canada.

However, if an expansively phrased restricted use clause in a confidentiality agreement effectively represents an area of exclusion clause of potentially unlimited geographic scope, what purpose is served by an actual area of exclusion clause in such a confidentiality agreement? Koenigsberg drew a subtle distinction between the two clauses. The area of exclusion provisions were held to represent an absolute prohibition against staking within the specified two-kilometre zone – whether or not the staking party made a discovery independently of the confidential information which was subject to the confidentiality agreement. By contrast, the restricted use provisions represented only a conditional prohibition in circumstances where the recipient had received confidential information and misused such information to stake the ground.[85]

The court in *Aquiline* also chose to explore in *obiter* whether IMA had breached a common law duty of confidence owed to Minera. Based on the tests set out in LAC Minerals, the court also determined that such a breach had occurred.[86]

Relying in part upon the *LAC Minerals* decision, the court concluded that an award of damages for IMA's breach of its confidentiality obligations would be inadequate in this case, due to the extreme difficulty in arriving at an assessment of the value of an exploration stage property.[87] Instead, the court issued a declaration that IMA held the Navidad Project pursuant to a constructive trust in favour of Minera. The court also granted a mandatory injunction requiring IMA to transfer the Navidad Project and any assets related thereto to Minera or its nominee, subject to the reimbursement of all reasonable amounts expended by IMA for the acquisition and development of the Navidad Project to date.[88]

The wide-ranging interpretation given to restricted use clauses in *Aquiline* was paralleled in a hostile takeover bid situation by the Ontario Superior Court of Justice in *Certicom Corp v Research In Motion Limited*.[89] In Certicom, the court issued an injunction against Research In Motion's (RIM) unsolicited takeover bid for Certicom. The court ruled that RIM's use of confidential information for the purpose of a hostile bid breached two non-disclosure agreements between

---

85 *Id* at 242.
86 *Id* at 244-253.
87 *Id* at 287.
88 *Id* at 292-293.
89 *Certicom Corp v Research In Motion Limited* (2009), 94 OR (3d) 511 (Ont Sup Ct J).

RIM and Certicom. Those agreements specifically restricted the use of confidential information by RIM to assessing a business relationship (including a business combination) between the parties – which was interpreted as excluding a hostile bid. The court also held that the restricted use clauses in question operated to prohibit a hostile bid independently of an express standstill clause between the parties – notwithstanding that the standstill clause had already expired. Accordingly, as a consequence of this ruling, the restricted use clauses of many run-of-the-mill confidentiality agreements in the mining industry may now inadvertently have the same practical effect in shielding targets from hostile bids as explicit standstill agreements. The reasoning in Certicom was later adopted by the Delaware courts in *Martin Marietta Materials, Inc v Vulcan Materials Co.*[90]

Receiving parties are becoming increasingly wary of broad restricted use clauses. Even prior to the *Aquiline* and *Certicom* decisions, Canadian commentators had noted that restricted use clauses presented receiving parties which have internal development activities in place which parallel the activities, or even the mere future plans and theories, of the disclosing party with "substantial and troubling questions about how a party receiving confidential information is ever to rid himself of the possibility that he will be accused later of utilizing that information to give himself a head start, or advantage, even though the information appears nowhere in his products or processes".[91]

Now, with the *Aquiline* and *Certicom* cases having demonstrated that restricted use clauses will be interpreted very broadly by Canadian courts and, unless explicit wording to the contrary is used, will trump more explicit prohibitions within an agreement (eg, area of exclusion and standstill clauses), receiving parties can be expected to take greater pains in negotiating restricted use provisions. Receiving parties may take the position that a disclosing party's reasonable expectations and concerns respecting potential misuse of its confidential information can be adequately addressed by specific disclosure restrictions, area of exclusion provisions, standstill clauses and other explicit, targeted prohibitions – without having to resort to an overbroad, catch-all restricted use clause. Other ways of limiting the unintended impact of restricted use clauses is for the receiving party to limit the extent of its diligence, reduce the term of the confidentiality agreement, adopt an internal firewall and sequester a 'clean team' to advance development opportunities without being contaminated by received confidential information;[92] or to rely upon a 'permitted use' clause, which is discussed below.

---

90 *Martin Marietta Materials, Inc v Vulcan Materials Co*, 56 A 3d 1072 (Del Ch 2012), *aff'd* 68 A 3d 1208 (Del 2012).

91 Richard A Brait, "The Unauthorized Use of Confidential Information" (1991) 18 *CBLJ* 323 at 349.

92 But see *Simplexity, LLC v Zeinfield*, 2013 WL 5702374 (Del Ch) (a firewall is not a cure-all, cannot be so general as to be meaningless and must be adhered to).

Permitted uses: The contractual flipside of the restricted use clause is the 'permitted use' clause. These clauses provide that confidential information may be freely used by a receiving party for a particular permitted purpose or permitted activity – such as competitive staking of mineral rights in a geographic region – regardless of the receiving party's possession of confidential information. Creative applications of permitted use clauses can often bridge the gap between disclosing and receiving parties faced with a negotiating impasse over the reasonable scope of a restrictive use clause, by ensuring that receiving parties will be permitted to proceed with specific internal development activities without subjecting those activities to the risk of future claims of breach of confidentiality based on the receipt of information which may overlap or touch upon such activities.

A permitted use clause was the subject of considerable judicial scrutiny in the British Columbia Supreme Court decision in *NovaWest Resources v Anglo American Exploration.*[93] The facts in this case were very similar to those in *Aquiline*. Novawest, a junior exploration company, sought to interest Anglo American Exploration, the exploration arm of a major gold producer, in jointly pursuing the exploration of mineral claims that Novawest had staked in the Cape Smith belt of northern Quebec. On 8 February 2002 the parties entered into a confidentiality agreement which contained a restricted used clause, pursuant to which Anglo would not "disclose any confidential information to any persons or parties or use any confidential information in competition with the Disclosing Party". The agreement also contained a one-kilometre area of exclusion clause.

On 7 and 19 February 2002 representatives of Novawest and Anglo held meetings, for which no written agendas, minutes or comprehensive notes were prepared by either party. At those meetings, Novawest alleged that its representatives showed confidential maps and documents to the representatives of Anglo and discussed Novawest's business plans (including a proposed expansion of staking activities into the West Raglan extension of the Cape Smith belt) and certain theories about the mineral deposits to be found in the Cape Smith belt. In their depositions, Anglo's representatives did not recall any information provided at such meetings as being confidential. Anglo also took the position at trial that it was unclear whether Novawest owned the information it claimed was confidential.

Preliminary discussions on the terms of an option and joint venture agreement were ongoing and continued into early May, but on 9 May 2002 Anglo advised Novawest that it would not be proceeding with further negotiations and that any data respecting the claims that Anglo had received from Novawest would be returned. Contemporaneously with breaking off its

93 *Supra*, note 42.

discussions with Novawest, over the period from 6 to 10 May 2002, Anglo staked a number of mineral claims in the West Raglan extension. All of these claims were between three and 70 kilometres west of the one-kilometre area of exclusion zone set out in the confidentiality agreement. Approximately two months following the staking, a colourful internal Novawest email was circulated which casts light on the mindset that some disclosing parties may bring to the exchange of confidential information:

> *WOW, have they covered the Wwweeeeeeessssstttttttt Eeeexxxxxttteeennn sssiiiooonnn!!!!!!!!!!! ...*
>
> *They were taking absolutely no chances and tied up the entire west extension of the Raglan main camp. We always said that it would be quite something to have a major do that and it's become a reality. Now we have to figure out how to use it to leverage off ...* [emphasis added][94]

Two years after the staking, Novawest commenced an action against Anglo alleging that Anglo's stakings in the West Raglan extension were based on confidential information supplied by Novawest at the meetings on 7 and 19 February and constituted a breach of the restricted use clause in the confidentiality agreement. Novawest sought the imposition of a constructive trust over the staked properties.

The crucial difference in facts between the *Novawest* and *Aquiline* cases, however, was that the confidentiality agreement between Novawest and Anglo contained the following permitted use clause:

> *8. NO RESTRICTION ON ACTIVITIES. It is hereby agreed that the Receiving Party's business activities include the evaluation and acquisition of mineral properties, and nothing contained herein will restrict or prevent it from acquiring any property through option, joint venture or staking of new mineral claims, except within a one kilometre area of influence from the current property boundary as shown on the attached Exhibit B (the "Area of Influence").*[95]

In light of the foregoing permitted use language, Justice Edwards characterised the nature and the commercial purpose of the confidentiality agreement between Novawest and Anglo in a very different light from that employed in the *Aquiline* decision:

> *Such agreements are for the benefit of both the disclosing party and the party receiving confidential information.*
>
> *The disclosing party benefits by having the receiving party evaluate claims without the risk that party will stake in the agreed area of influence around those claims as a result of confidential information disclosed.*
>
> *The receiving party is able to consider confidential information without subjecting*

---

94 *Id* at [33] (on examination for discovery, Novawest's representative described this email as being "facetious").

95 *Id* at [17].

subsequent staking it does outside the agreed area of influence to claims of breach of confidentiality on the basis that information disclosed by the disclosing party led to that staking.

. . .

I find the commercial purpose of the Confidentiality Agreement was to avoid disputes about which party had what information and what information could be used in deciding to stake outside the area of influence. Its terms were required by Anglo in order to entertain Novawest's "pitch" respecting the Novawest claims, so as to avoid the very dispute which has now arisen.[96]

The court held that the plain meaning of the permitted use language in Clause 8 allowed Anglo to stake mineral claims outside the area of exclusion. The court also determined that the permitted use language in Clause 8 supplanted any common law duty of confidentiality that Anglo owed to Novawest.[97] Edwards was careful to distinguish the circumstances of the relationship between Novawest and Anglo from the relationship between International Corona and LAC Minerals which gave rise to a common law duty of confidence in the *LAC Minerals* decision:

Counsel agreed that a consequence of LAC was the adoption in the mining industry of confidentiality agreements such as the one in this case. There was no such agreement between the parties in LAC. La Forest J. commented at para. 661 of LAC: "When the parties have reduced their understandings to writing, it is obviously the proper course for courts to be extremely circumspect in adding to the bargain they have set down."[98]

Although the court in *Novawest* reached a decision which, on its face, might seem to be the opposite of that reached in *Aquiline*, the two cases can be reconciled on the basis that the permitted use clause in *Novawest* clearly granted the defendant the right to stake mineral claims outside the area of exclusion. Canadian courts will treat clearly drafted expressions of contractual intent with circumspection and will be reticent to impose a common law duty of confidence that flies in the face of such clear contractual provisions. By contrast, in *Aquiline* there was no such clear permitted use clause. In the absence of clear contractual language that preserves the ability of the receiving party to undertake specific activities irrespective of its receipt of confidential information, Canadian courts will likely uphold broad restricted use clauses and invoke the common law duty of confidence to prevent receiving parties from misuse of confidential information.

Standstills: Due in part to the robust level of M&A activity within the mining industry over the past several years (and particularly the noteworthy increase of

96 *Id* at [60] – [62] and [89].
97 *Id* at [91].
98 *Id* at [59].

hostile M&A activity during that period), disclosing parties that are publicly traded companies will frequently seek to include a standstill clause within their confidentiality agreements. Such clauses broadly prohibit:

- acquisitions of the disclosing party's securities;
- mergers or other business combinations with the disclosing party;
- solicitations of proxies to vote the disclosing party's securities;
- participation in a group in respect of any voting securities of the disclosing party;
- an attempt to influence or control the disclosing party;
- any discussions, negotiations, arrangements or understandings with respect to any of the foregoing; and
- the provision of advice or assistance, or acting as a financing source, with respect to any of the foregoing.

The goals of standstill provisions are to prevent friendly discussions from turning into hostile transactions which may put the disclosing party into play and to provide the disclosing party's board of directors with some measure of control over the ability of receiving parties to acquire shares of the disclosing party. Although open to challenge, the use of standstill covenants will not generally, in and of itself, be held invalid in a competitive auction process for a public company in Canada or the United States.[99]

Receiving parties may be amenable to the inclusion of standstill provisions in a confidentiality agreement where the confidential information that has been disclosed is material to the disclosing party and places the receiving party in a position to take advantage of the disclosing party by purchasing its stock or engaging in other hostile conduct intended to influence control. Accordingly, receiving parties will generally be more willing to agree to standstill clauses in circumstances where the confidential information disclosed is not merely related to exploration results from a single mineral property, but includes material information pertaining to all of the disclosing party's material properties, as well as material corporate and financial data on information pertaining to the disclosing party's existing commercial relationships. However, where the disclosing party is a junior exploration company with only a single significant property, material undisclosed exploration results from that property may well be the most material information pertaining to the disclosing party and justify the protections of a standstill clause.

One of the most debated aspects of standstill clauses is their duration. Disclosing parties will seek to obtain standstill protection for as long as possible, often matching the duration of the confidentiality provisions of a

99 See *Rogers Communications Inc v Maclean Hunter Ltd* (1994), 2 CCLS 233 (Ont Ct (Gen Div)) at 248; *Re JP Stevens & Co Shareholders Litigation* [1987-8] Fed Sec Law Rep 98,377 at 98,387 (Del Ch Ct).

confidentiality agreement (eg, two to five years). From the receiving party's standpoint, however, a standstill clause is necessary to protect the disclosing party only where the receiving party has been provided with undisclosed information of sufficient materiality that the receiving party has an advantage over the marketplace generally in valuing the disclosing party and its publicly traded securities. If the confidential information provided to the receiving party truly rises to this level of materiality, then the disclosing party will likely be required to make such information public in the very near term in order to comply with securities laws. Once such information is made public, however, the rationale underlying the disclosing party's need for standstill protection will have fallen away – and the receiving party will not want to be sidelined in those circumstances from participating in an auction for control over the disclosing party if it is put into play. Accordingly, receiving parties will be unlikely to accept standstill provisions with a duration of two years or longer, and terms of six to 18 months will be more acceptable.

One important issue to consider in connection with standstill clauses is the extent to which the standstill provisions are conditional upon material confidential information actually being provided to the party agreeing to the standstill. This issue was the subject of the decision of the British Columbia Supreme Court in *Aurizon Mines Ltd v Northgate Minerals Corp*,[100] which was subsequently upheld upon appeal to the British Columbia Court of Appeal.

In October 2005 Aurizon Mines Ltd and Northgate Minerals Corp entered into discussions concerning a possible business combination. Each party entered into separate confidentiality agreements with the other, but only Northgate was subject to a standstill clause. The wording of this standstill clause was in line with similar clauses used within the industry.

On 15 December 2005, before any confidential information was actually exchanged, Aurizon advised Northgate that it no longer wished to pursue a transaction or continue discussions with Northgate. Northgate reasoned that since no confidential information had been exchanged between the parties, the confidentiality agreements and the parties' obligations thereunder (including the standstill clause) were at an end. Northgate advised Aurizon of its position both orally and in a letter on 19 December 2005. Aurizon did not respond to Northgate's letter.

In February 2006 Northgate began purchasing voting securities of Aurizon and in May 2006 it began communicating with Aurizon's shareholders regarding an acquisition of Aurizon's voting securities. On 23 May 2006 Northgate announced its intention to make an unsolicited offer – a euphemism for a hostile takeover bid – for all of Aurizon's outstanding voting securities. On

100 *Aurizon Mines Ltd v Northgate Minerals Corp* (2006), 57 BCLR (4th) 137 (BCSC), *aff'd*, (2006), 55 BCLR (4th) 203 (BCCA).

1 June 2006 Northgate's formal offer was launched. Aurizon sought an injunction against Northgate's bid, asserting that the standstill provisions contained in Northgate's confidentiality agreement were still effective and that Northgate was in breach of those provisions

The British Columbia Supreme Court's interpretation of the standstill clause in *Aurizon* was ultimately very similar to the court's interpretation of the area of exclusion clause considered in *Aquiline*. Justice Allan held that it was not an express term of Northgate's confidentiality agreement that the standstill provisions would apply only if confidential information were exchanged between Northgate and Aurizon.[101] Neither could such a term be implied into the confidentiality agreement. Furthermore, the court determined that the confidentiality agreement in question took effect upon its execution, not upon the date that confidential information was first provided. Accordingly, the court declared that the terms of Northgate's confidentiality agreement remained binding on Northgate until its expiry on 19 October 2006 and Northgate was enjoined from pursuing its offer until that date.[102]

In light of the decision in *Aurizon*, it is reasonable – and prudent – for receiving parties to insist that the enforceability of any standstill provision is conditional upon material information pertaining to the disclosing party (possibly identified in a schedule to the confidentiality agreement) actually having been disclosed to the receiving party.

Standstill clauses which are too restrictive can at times backfire against a public mining company engaged in an auction process and prevent higher offers. This is because most friendly support or arrangement agreements require the target to covenant not to release any third party from, waive or forbear in the enforcement of, any confidentiality or standstill agreement. In *Ventas Inc v Sunrise Living REIT*, the Ontario Superior Court, in a decision upheld by the Ontario Court of Appeal, strictly enforced such a covenant by Sunrise not to waive the standstill provision in any of its confidentiality agreements with other parties – notwithstanding that the purchase agreement contained a 'fiduciary out' allowing Sunrise to consider *bona fide* unsolicited superior proposals.[103] As a result, Sunrise was obligated to enforce a standstill with a third party, even though such third party had subsequently made a superior offer to acquire Sunrise.

To avoid *Ventas*-type situations, many standstill clauses are subject to a 'spring' clause, under which the standstill will terminate if a control transaction (friendly or unsolicited) is publicly announced. Canadian courts have upheld such spring clauses.[104]

101 *Id* at 150-151.
102 *Id* at 154.
103 (2007), 29 BLR (4th) 292 (Ont Sup Ct J), *aff'd* (2007), 85 OR (3d) 254 (Ont CA).
104 *Quebecor Media Inc v Osprey Media Income Fund, Ont* [2007] OJ No 3070 (Ont Sup Ct J).

Receiving parties may also insist that a 'most favoured nation' clause be included within the standstill provisions. Such clauses operate to reduce the duration and scope of the standstill protections if the disclosing party enters into another standstill agreement with a third party that is of lesser scope or duration than the original standstill provisions. Another frequent exclusion to standstill clauses is a 'toehold' exclusion, which permits the recipient of information to acquire a small stake in the disclosing issuer (2% to 5%). Such a 'toehold' exclusion may reduce the receiving party's risk of being unable to successfully conclude a transaction by allowing it to recoup at least some of its costs by selling the toehold shares to a superior competing offer.

(f) Areas of exclusion

Confidentiality agreements in the mining industry commonly contain 'area of exclusion' clauses (also referred to as 'area of interest' or 'area of mutual interest' clauses), which prohibit recipients of confidential information from acquiring mineral or surface rights within a defined area surrounding the outer perimeter of a disclosing party's mineral or surface rights for a defined period.[105] Such clauses effectively prohibit receiving parties from using a disclosing party's confidential geological data to engage in competitive acquisitions of mineral rights in immediately adjacent areas. Often these prohibitions on the acquisition of mineral and surface rights are absolute and do not turn on any improper use of confidential information by a receiving party as a basis for such acquisitions in the excluded area.[106] Some broader area of exclusion clauses will prohibit not only acquisitions within the area, but also any exploration, development or mine operation activities. Parties should also consider whether an area of exclusion clause should extend to the acquisition of water rights in the area.

Restrictions on the size of such areas of exclusion can vary from hundreds of metres to five or more kilometres distant from the outer perimeter of the property to which the confidential information pertains. Their duration can last for up to several years.

Area of exclusion clauses can readily be justified by disclosing parties where:

- mineral or surface rights surrounding or adjacent to the outer perimeter of the disclosing party's property remain available for staking (ie, 'open' ground);
- the receiving party has not, or has only recently, initiated regional exploration activity in the immediate vicinity of the disclosing party's property; or
- the confidential information disclosed has relevance to the evaluation of

---

105 See, for example, Form 7, *supra* note 78 at Appendix 5.
106 See, for example, *Aquiline*, *supra* note 37 at 234.

the mineral character or value of mineral or surface rights in the immediate vicinity of the disclosing party's property.

However, less justification exists for area of exclusion clauses where:

- mineral or surface rights surrounding or adjacent to the outer perimeter of the disclosing party's property are not available for staking;
- the receiving party already has an active regional exploration programme ongoing in the vicinity of the disclosing party's property; or
- the geological features of the region suggest a lack of continuity between the disclosing party's property and other mineral deposits potentially in the vicinity of such property.

Receiving parties will often seek to limit area of exclusion clauses in confidentiality agreements to a prohibition against the receiving party staking adjacent open ground, rather than acquiring adjacent mineral or surface rights already held by third parties. The rationale behind this position is that in the absence of the receiving and disclosing party entering into a joint venture, a receiving party should be free to entertain offers from third parties holding ground in the region and should certainly not be restricted from engaging in a corporate merger or control transaction with a company that also holds ground in the region.

As noted above, in *Aquiline* a typical area of exclusion clause was held to be an "independent covenant",[107] much like the similarly independent standstill covenant in *Aurizon*, and enforceable even if no confidential information had been provided to the receiving party concerning the disclosing party or its property. Accordingly, it is reasonable – and likely prudent, in the wake of the *Aurizon* decision – for receiving parties to insist that the enforceability of any area of exclusion provisions of a confidentiality agreement be conditional upon material information pertaining to the disclosing party's property (possibly identified in a schedule to the confidentiality agreement) having been disclosed to the receiving party.

As discussed earlier, *Aquiline* also established that, unless explicit language to the contrary is employed in a confidentiality agreement, a typical independent area of exclusion clause will not prevent a restricted use clause or the common law duty of confidence from being applied to prohibit a receiving party from staking ground outside of the boundaries of the specified area of exclusion – provided that such staking results from the misuse of confidential information.[108] In the wake of *Aquiline* and *Novawest*, it would be prudent for receiving parties to insist upon the inclusion of a permitted use clause in

107 *Supra* note 37 at 240 (BCSC).
108 *Id.*

confidentiality agreements – similar to that employed in *Novawest* – which explicitly allows the receiving party to stake or otherwise acquire ground outside the boundaries of any area of exclusion.

Area of exclusion clauses in confidentiality agreement are subject to many, if not all, of the myriad drafting considerations that are applicable to area of interest provisions in an option or joint venture agreement, which are beyond the scope of this chapter.[109]

(g) Non-solicitation

One of the most valuable assets of any mining company is its employees. As transactions involving the exchange of confidential information from one mining industry participant to another will generally involve contact between the respective employees of the parties, disclosing parties will often be concerned that a receiving party may – either through the information being provided or simply through a relationship which develops over the course of the due diligence process – learn of the strengths of the disclosing party's employees and recruit them. In addition, hiring away the employees of a disclosing party may be an effective indirect method of acquiring confidential information in respect of a disclosing party's properties. Accordingly, confidentiality agreements frequently include a covenant from the receiving party that it will not employ or directly solicit employees of the disclosing party.[110] Such non-solicitation clauses are necessitated by the unwillingness of Canadian courts to extend the common law duty of confidence to restrict the solicitation of employees.[111]

As non-solicitation clauses are effectively a variety of non-compete covenant, their scope should be carefully limited by disclosing parties in order to ensure that these clauses are not held invalid for public policy reasons. However, Canadian courts are more likely to uphold non-solicitation restrictions in the context of a sale, relative to the employment context.[112]

Receiving parties often try to restrict non-solicitation clauses to active solicitation of the disclosing party's employees, and exclude the hiring of individuals who apply for positions of their own volition. Receiving parties will also try to restrict the categories of a disclosing party's employees who are subject to a non-solicitation clause to those specific employees who participated

---

109 See Justin J Quigley and Stephen Hull, "Give What? To Whom? But Why? Area of Interest Provisions in Mining Agreements" (2009) 55 *Rocky Mt Min L Inst* 38-1; Fred R Pletcher, "Preferential Purchase Right and Area of Interest Clauses" in *Mining: Drafting Joint Venture Agreements* (1999); WF Harasym, "Area of Interest Provisions in Mining Agreements" in *Mining Agreements III* (1991) 11-1; MT Nesbitt, "Area of Interest Provisions - Two Edged Swords" (1989) 35 *Rocky Mt Min L Inst* 21-1; Zarlengo, "Area of Mutual Interest Clauses Regarding Oil & Gas Properties: Analysis, Drafting, and Procedure" (1982) *28 Rocky Mt Min L Inst* 837; LD Clark, "Area of Interest Provisions" in *Mining Agreements II* (1981) 601.

110 See, for example, Form 7, *supra* note 44 at Appendix 4.

111 *Chevron Standard Ltd v Home Oil Co* (1980), 22 AR 451 at 493-96 (QB), aff'd (1982), 19 Alta LR (2d) 1 (CA) (the possibility of misuse of confidential information by a former employee is not a sufficient threshold for the former employer to obtain legal redress).

112 *Payette v Guay Inc* [2013] SCJ No 45; *Shafron v KRG Insurance Brokers (Western) Inc* [2009] 1 SCR.

in the due diligence process and possess material scientific and technical knowledge concerning the property or properties for which confidential information has been disclosed to the receiving party. Otherwise, routinely entering into such clauses can greatly complicate a receiving party's ordinary course recruitment efforts. Some receiving parties may even insist that a non-solicitation clause be a two-way covenant, so that both parties' due diligence personnel are protected.

(h) Return or destruction of data

Confidentiality agreements will usually provide that, upon request by the disclosing party, the receiving party will return or destroy all confidential information, including copies of that information.[113]

Some consideration should be given as to whether an obligation to return or destroy data should arise only upon receipt of a written request from the disclosing party, or whether it should arise automatically once one party has informed the other that it will not be proceeding with a proposed transaction. The latter provision will ensure that a disclosing party's confidential information does not remain in the hands of a competitor simply as a result of an inadvertent failure to give notice. Had the confidentiality agreement among Newmont, Minera and IMA contained an 'automatic' return clause, for example, the dispute between the parties would not likely even have arisen (as in that case neither Newmont nor Minera ever followed up with IMA to request the return or destruction of confidential information. However, an automatic requirement to return or destroy the information would likely have prevented IMA's geologist from examining the critical BLEG A data after IMA had determined not to bid on the Calcatreu Project).

Receiving parties will wish to ensure that such 'return or destruction' clauses have clear and workable timeframes, and that such timeframes are adhered to in practice. Otherwise, if the receiving party has inadvertently breached a covenant to return or destroy confidential data, it may not be in a position to rely upon the termination clause of the confidentiality agreement – which will likely extend the lifespan of any non-disclosure, restricted use, standstill and area of exclusion provisions.

Disclosing parties may wish to ensure that return or destruction clauses extend to all 'derivative works', such as technical and interpretative reports, data and other materials generated by the receiving party from the disclosed information. In such circumstances, receiving parties may want to ensure that the receiving party has the option to destroy, rather than return, such derivative works. Otherwise, the receiving party will have expanded the disclosing party's database, without compensation.

---

113 See, for example, Form 7, *supra* note 44 at §6.

A receiving party may seek an exception to return or destruction provisions for interpretative information that has been considered by its board of directors and forms part of the minutes of its meetings or where, in the opinion of the receiving party's counsel, such return or destruction would be unlawful or violate any order, judgment, writ or decree by which the receiving party is bound. In addition, a receiving party may seek the right to retain a single copy of all confidential information (perhaps stored in a secure file offsite or with the receiving party's counsel), solely for evidentiary purposes in case of some future dispute. Otherwise, the return or destruction of such information may place the receiving party at a considerable evidentiary disadvantage in any future litigation when it comes to proving what was or was not disclosed.

Where confidential information is provided in electronic format, consideration should be given to address the practicalities of returning or destroying such information if it becomes part of the receiving parties' back-up or archived electronic records. Exceptions to return or destruction provisions for back-up or archived electronic records are increasingly common.

Finally, consider whether the confidentiality agreement should require the receiving party to provide the disclosing party with an officer's certificate attesting that the confidential information has been returned or destroyed in accordance with the agreement.

(i) *Relief*

As it may be difficult to establish damages that result from a breach of confidentiality, it is useful to provide an acknowledgement by the receiving party in a confidentiality agreement that the disclosing party would suffer irreparable harm upon a breach and, thus, the disclosing party should be entitled to seek an injunction restraining the receiving party from breaching the provisions of the agreement as a remedy in a successful action.[114] Often, these 'relief' clauses will waive the requirement for the disclosing party to meet certain civil procedures or evidentiary requirements, such as security for costs or establishing that damages are inadequate and cannot be determined. However, receiving parties must be careful to ensure that such relief clauses are not drafted so broadly that they eliminate the need for the disclosing party to meet the other evidentiary burdens ordinarily required to obtain an injunction.

The parties should be aware that courts will not automatically give effect to such relief provisions.[115] Rather, the courts will generally view such clauses as mere evidence that the traditional common law remedy of damages is regarded by the parties as inadequate. Because relief clauses are often considered to

---

114 Thorburn, *supra* note 2 at 7:3210 and RP Lynch, *The Practical Guide to Joint Ventures and Corporate Alliances*, 1st ed (New York: John Wiley & Sons, 1990) at 151.

115 See *Brenner (Trustee of) v Brenner* [1923] 3 DLR 1097 (Ont SC).

constitute evidence of the parties' intentions, it is useful to set out within such clauses the rationale for including them within the confidentiality agreement (eg, because the confidential information is the product of the expenditure of considerable time, effort and money).[116] Regardless, in most cases involving a breach of confidence, an injunction to restrain unauthorised disclosure of confidential information is likely to be granted.[117]

Disclosing parties should also be aware that adverse legal consequences can flow from overstating the inadequacy of damages in a relief clause. In certain circumstances, it may be appropriate for a disclosing party to bring an action for damages (eg, where the receiving party may have a defence such as laches). In such circumstances, the presence of a clause in the confidentiality agreement which states that damages "cannot be calculated or determined" in the event of a breach will likely prove an obstacle to recovery. Consequently, relief clauses should be carefully drafted to reflect that damages are inadequate, not that compensation by way of an award in damages is impossible.[118]

4. Conclusion

The profound financial impact of the disputes underlying many of the leading cases relating to confidentiality in the Canadian mining industry, such as *LAC Minerals* and *Aquiline*, serve as a strong reminder that the use and management of confidential information in mining industry transactions can be critical. Counsel should recognise that where a significant exploration discovery is made on a property that is subject to a confidentiality agreement, that agreement will likely be scrutinised in painstaking and minute detail by the litigation attorneys of the other parties to the agreement (and the other parties' former partners and so on), to determine whether they can – in the words of the damning email from *Novawest* – "figure out how to use it to leverage off" and serendipitously profit from the receiving party's exploration or development success, without incurring significant exploration or development risk. However, if parties are careful to tailor the language of confidentiality agreements to reflect the particular type of transaction, the specific confidential information being provided and the reasonable underlying objectives of the parties, costly and time-consuming negotiation and posturing over these agreements can be minimised and unintended *LAC Minerals* and *Aquiline*-type results can perhaps be avoided. Hopefully this chapter has illustrated some of the considerations that should be taken into account in the preparation of such agreements.

---

116 CI Kyer, "Contracting for Injunctive Relief: Is it Effective?" (1986), 3 CCLR 165 at 167-68.
117 *Id* at 167-68.
118 *Id* at 167-68.

Preliminary agreements

Dawn G Meidinger
Sarah A Strunk
Fennemore Craig, PC

Preliminary agreements are a routine step in the development of many exploration prospects and mineral development properties around the world. These early stage agreements are often entered into by international exploration professionals without the assistance of lawyers in order to set down the basic terms of the transaction to explore, develop, finance, acquire or divest in a mineral project. Because the terms and relationships in mineral development can be very complicated, the preliminary agreement is typically the first step in documenting the basic terms of the transaction between a mineral owner and a party interested in exploring and developing the project. These agreements are referred to by a variety of names, including 'heads of agreement', 'agreement to negotiate', 'memorandum of understanding', 'letter of comfort', 'agreement in principle', 'memorandum of agreement' or 'letter of intent'. While this chapter focuses on US legal concepts, the discussion and considerations are relevant in many international jurisdictions where mineral exploration and development take place.

1. What are preliminary agreements?

In essence, a preliminary agreement is any general understanding between parties regarding a contractual arrangement where the parties anticipate that they will draft, negotiate and execute a definitive and binding agreement incorporating the substance of the preliminary agreement. Preliminary agreements can be short and summary or lengthy and detailed, depending on the preference of the parties.[1]

2. Why do parties utilise preliminary agreements?

Parties enter into preliminary agreements for a variety of reasons. For example, they may lack sufficient information to fully commit to a transaction at the time; or they may want to outline the general terms of a deal and then conduct additional due diligence or technical evaluations to determine whether they will enter into a definitive agreement addressing all necessary terms and conditions.

1 Preliminary agreements can also be oral, but this chapter explores only written agreements.

Preliminary agreements focus the parties on identifying and resolving material issues and keep both parties on track while negotiating the remaining terms and conditions necessary for a binding contract. They are particularly helpful in complex mineral exploration and development transactions, where parties enter into preliminary agreements to memorialise basic terms and conditions to facilitate further negotiations.[2]

Issues of expediency also drive the use of preliminary agreements. For example, if investment or financing is needed to complete a transaction, a preliminary agreement with sufficient detail may be used to attract the necessary capital and start the negotiations on terms for such investment and financing for closing. Likewise, if a transaction must be closed quickly and there are regulatory requirements, such as pre-merger notification requirements, a preliminary agreement can be used to satisfy governmental filing requirements. A preliminary agreement allows the parties to begin the various regulatory reviews earlier, while negotiating the definitive agreements, and may also form the basis for some public disclosure about the transaction.

Even if the terms of a preliminary agreement are generally non-binding, the preliminary agreement may provide binding exclusivity for a period while due diligence is being conducted. Similarly, one party may view a preliminary agreement as a way to gauge whether the other is serious about a transaction. These reasons can drive a decision to enter into a preliminary agreement. Hiring consultants such as environmental firms and professional land/title experts, engaging lawyers to draft definitive documents and generally engaging in due diligence is expensive. Before a mineral developer elects to expend such funds, it will most likely want to know that it is not competing with other developers or pursuing a transaction that is wholly unacceptable.

It can be perilous, however, if the binding nature of the preliminary agreement is not expressly stated. For example, if the desire to enter into a preliminary agreement is to facilitate confirmation drilling prior to entering into a full agreement, the results of the drilling may colour the parties' thinking about whether the preliminary agreement is binding. If the results are positive, the party seeking to explore, develop, earn an interest in or purchase the mining project may assert that the agreement is binding; while a farming out or selling party may contend that the preliminary agreement is not binding. Dispute resolution associated with this situation will likely be more costly than taking the time to negotiate a definitive agreement.

There is a common assumption that preliminary agreements are less expensive than negotiating complex definitive agreements. This assumption is usually correct, but other considerations must be taken into account.

---

2 See George Reeves, "Preliminary Agreements", *Mining Agreements III* (Mineral Law Series vol 1991, no 3, May 1991).

3. Reasons to avoid using a preliminary agreement

The principal reason to avoid using a preliminary agreement is that the agreement may not be binding and enforceable as to the material business matters. Expressly non-binding agreements can be discarded at any time and the material terms are not enforceable. While a non-binding preliminary agreement might provide the parties with flexibility, the associated legal risk must be considered.

Confidentiality and public disclosure provisions often need to be binding and should not be overlooked when entering into a preliminary agreement.[3] If pre-closing confidentiality about the subject matter of a preliminary agreement is desired, consideration should be given to the public disclosure requirements that may result from its execution. A preliminary agreement may trigger a public disclosure that one side of the transaction may not desire.[4]

The often cursory nature of preliminary agreements can be problematic if the agreement is found to be binding and enforceable. If a preliminary agreement fails to fully explore all aspects of the transaction intended by a party – even if it is by reference to other documents such as forms that have been drafted for the mineral industry – it could be detrimental to have the preliminary agreement deemed binding and enforceable.

Another reason to forgo a preliminary agreement is that the drafting and negotiation will delay and add to the overall cost of an exploration or mineral transaction. This is especially true where there is a clear and sufficient understanding of the material terms of the mineral exploration or development transaction that makes use of a preliminary agreement an unnecessary expense or delay. If a mineral transaction must be closed quickly, the parties should consider moving towards a final and definitive agreement rather than incurring the delay of negotiating a preliminary agreement.

Client and counsel should discuss what they hope to accomplish by using a preliminary agreement and its potential consequences. Key inquiries include the following:

- Do the parties intend to create additional obligations or are they merely expressing an interest to negotiate further?
- What expectations do the parties have of each other?
- Do the parties intend to be bound by the terms of the preliminary agreement and if so, to what extent?

4. Enforceability of preliminary agreements

Preliminary agreements can be binding, partially binding or non-binding. Before drafting, parties must contemplate the consequences of their actions and

3 See generally, Caroline Thomas and Carmen Diges, *Model for a "Modern Major-General": Key Negotiating Considerations and Legal Implications for Letters of Intent and Option Agreements in Exploration Deals,* Rocky Mountain Mineral Law Institute, Fifty-Eighth Annual Institute, 2012, §28.02.

4 *Id.*

clearly state their intent or risk enforceability issues. The question of whether a preliminary agreement is enforceable has been the subject of substantial litigation, as examined below.

4.1 *Teachers Insurance*

In the United States, an often-cited case involving a preliminary agreement is *Teachers Insurance and Annuity Association of America v Tribune Company*, 670 F Supp 491 (SDNY 1987). The case involved a dispute between a lender and a borrower surrounding an alleged breach of a commitment agreement for a loan. The letters constituting the commitment agreement stated that there was a binding agreement subject to the preparation and execution of final agreements satisfactory to both sides, and the approval of Tribune's board of directors as the borrower. Prior to the preparation of the final documents, Tribune broke off negotiations and declined to negotiate further unless the lender made additional concessions. When Tribune concluded a loan with another lender, the Teachers Insurance and Annuity Association of America sued.

The borrower contended that:

- although the commitment letter left many terms open and was an undertaking to negotiate, it did not obligate either side to enter into a loan that was adverse to its interest;
- both sides had reserved the right to approve satisfactory documentation; and
- its obligation was conditioned on the approval of its board of directors.[5]

The lender argued that although the commitment letter did not constitute a concluded loan agreement, it was nonetheless a binding commitment which obligated both sides to negotiate in good faith towards a final contract conforming to the agreed terms, and it thus committed both sides not to abandon the deal. The lender argued that the open terms were of minor economic significance and were covered by the provisions that referred to "customary and usual" terms in this type of transaction.[6]

The court in *Teachers* observed that:

Preliminary contracts with binding force can be of at least two distinct types. One occurs when the parties have reached complete agreement (including the agreement to be bound) on all the issues perceived to require negotiation. Such an agreement is preliminary only in form—only in the sense that the parties desire a more elaborate formalization of the agreement. The second stage is not necessary; it is merely considered desirable.

– – –

5 *Teachers*, 670 F Supp at 496.
6 *Id.*

> *The second and different sort of preliminary binding agreement is one that expresses mutual commitment to a contract on agreed major terms, while recognizing the existence of open terms that remain to be negotiated. Although the existence of open terms generally suggests that binding agreement has not been reached, that is not necessarily so. For the parties can bind themselves to a concededly incomplete agreement in the sense that they accept a mutual commitment to negotiate together in good faith in an effort to reach final agreement within the scope that has been settled in the preliminary agreement. To differentiate this sort of preliminary agreement from the first, it might be referred to as a binding preliminary commitment. Its binding obligations are of a different order than those which arise out of the first type discussed above. The first type binds both sides to their ultimate contractual objective in recognition that that contract has been reached, despite the anticipation of further formalities. The second type—the binding preliminary commitment—does not commit the parties to their ultimate contractual objective but rather to the obligation to negotiate the open issues in good faith in an attempt to reach the alternate objective within the agreed framework. In the first type, a party may lawfully demand performance of the transaction even if no further steps have been taken following the making of the "preliminary" agreement. In the second type, he may not. What he may demand, however, is that his counterparty negotiate the open terms in good faith toward a final contract incorporating the agreed terms.*[7]

The court in *Teachers* thus recognised two types of preliminary agreements. The first type of preliminary agreement ('Type I') is a binding agreement where the parties fully agree to all of the material and essential terms and the contract is a mere formality. Type I agreements are fully enforceable, even if the formal agreement is never executed. The second type of preliminary agreement ('Type II') is a binding agreement only to a certain extent, anticipating that there will be further negotiations of the open terms even if the major terms have been concluded. Type II agreements do not fully commit the parties to the objectives of the preliminary agreement, but bind the parties to negotiate in good faith to reach the objective of that agreement. If the parties are unable to resolve the open terms and execute a definitive agreement notwithstanding their good-faith efforts, neither party is bound to the business deal set forth in the preliminary agreement.

The court in *Teachers* found that the first and most important factor in determining which type of agreement is involved is the express language of the preliminary agreement. Specifically, the court evaluated whether the parties considered the preliminary agreement binding or whether they intended not to be bound until the conclusion of another, more formal agreement.[8]

---

7 *Teachers*, 670 F Supp at 498.
8 *Teachers*, 670 F Supp at 499.

The fundamental underpinning of contract law in most common law jurisdictions is to protect the intentions of the parties, no matter how informal their arrangements might be.[9] "A primary concern for courts in such disputes is to avoid trapping parties in surprise contractual obligations that they never intended."[10] Thus, any preliminary agreement should clearly state which provisions the parties intend to be binding, if any. If the drafter does this properly, most of the legal issues surrounding enforceability fade away. If the parties are unclear about their intent, then a court or other tribunal is left with the job of determining what the parties truly intended.

The court in *Teachers* did not end its analysis with the express intent of the parties. The court reviewed a number of other factors, including:

- the context of the negotiations of the parties;
- the existence of partial performance by one or more of the parties;
- what terms were 'open' or remained to be negotiated; and
- whether it was customary for the parties to reduce this type of transaction to a formal contract before they considered it binding.[11]

Each of these factors also requires a fact-intensive evaluation. While the preliminary agreement may expressly state that it is non-binding until the occurrence of some other event, these other factors may influence a court to determine a preliminary agreement as binding.

4.2 Expressing intent

Assuming that a preliminary agreement contains all the essential terms[12] and expressly states that the parties intend it to be a binding agreement or contract, courts will enforce the parties' intent regardless of whether the parties contemplated entering into another agreement.[13] If the parties do not intend to be bound by the preliminary agreement, they are advised to expressly include a provision similar to the following example:

THIS LETTER EXPRESSES THE INTENT OF THE PARTIES FOR DISCUSSION PURPOSES ONLY FOR USE IN DRAFTING A DEFINITIVE CONTRACT. THIS LETTER IS NOT INTENDED TO CREATE NOR SHOULD IT BE CONSTRUED AS CREATING ANY LEGAL OBLIGATION TO CONCLUDE THIS TRANSACTION UNDER THE TERMS OUTLINED HEREIN OR ON ANY OTHER TERMS OR CONDITIONS NOR IS IT INTENDED TO CREATE ANY OTHER

---

9 See *Winston v Mediafare Entertainment Corp* 777 F 2d 78 (2d Cir 1985); *RG Group v Horn & Hardart Co*, 751 F 2d 69 (2d Cir 1984); *Reprosystem, BV v SCM Corp*, 727 F 2d 257 (2d Cir), cert denied, 469 US 828, 105 S Ct 110, 83 L Ed 2d 54 (1984).

10 *Teachers*, 670 F Supp at 497.

11 *Teachers*, 670 F Supp at 500-503.

12 Even if the preliminary agreement states that it is intended to be binding, if it fails to demonstrate agreement on the material terms of a bargain, there can be no agreement. The essential terms of agreement must be present and cannot be substituted by the courts.

13 Arthur L Corbin, *Corbin on Contracts* §30 at p111 (1963).

OBLIGATION EXCEPT FOR THE OBLIGATIONS SET FORTH IN PARAGRAPHS __ AND __.[14]

Even when parties do not intend to be bound by the terms of a preliminary agreement, they may wish to make certain provisions binding, such as:

- an obligation to maintain confidentiality;
- rights to conduct due diligence or exploration activity;
- exclusivity of the negotiations; and
- public announcements or press releases.

In such case, use of the provisions that follow or similar language in the preliminary agreement may be helpful:

This Letter is not a binding agreement except to the extent specifically stated below. According to its terms, only three provisions were binding, to wit: the due diligence indemnification provision, the non-shop provision, and the deposit provision, leaving, inter alia, the provisions pertaining to alterations, the guaranty, and the right of first refusal open for negotiation.

– – –

The provisions of paragraphs __, __, __ and ___ (the "Binding Provisions") are all intended to be binding on the Parties. Otherwise, this [Heads of Agreement] is an expression of intent only and not intended to be a binding agreement between the Parties with respect to the remaining subject matter. A binding agreement between the Parties will not occur unless and until the execution of the Agreement. Until the execution of the Agreement, each Party shall have the absolute right to terminate all negotiations and this HOA, for any reason, without liability therefore save liability for any breach of a Binding Provision occurring prior to such termination.

– – –

This Heads of Agreement ("HOA") sets out the general terms and conditions of the transaction. This HOA is intended to focus and confirm our discussions and serve as the basis for negotiating and executing a definitive agreement or agreements documenting the Transaction (collectively the "Agreement"). Upon execution of this HOA, the Parties will proceed with good faith negotiation of the Agreement and make all reasonable efforts to effect the execution of the Agreement by ________ (the "Execution Date"). Except for the legally binding provisions set forth in this HOA, there is and will be no legally binding agreement between the Parties in respect of the Transaction until the execution of the Agreement.

Notably, each of the foregoing contractual provisions contemplates that there will be a subsequent definitive agreement. As previously discussed, making the subject of a preliminary agreement binding 'subject to' the execution of a

14 The court reviewing this clause concluded that the provisions of the letter unambiguously stated that the parties did not intend to be bound by the letter's terms. *White Const Co, Inc v Martin Marietta Materials, Inc*, 633 F Supp 2d 1302, 68 UCC Rep Serv 2d 832 (MD Fla 2009).

definitive agreement is sufficient for some courts to find that the preliminary agreement is not a Type I, fully binding and enforceable commitment. "When 'parties contemplate the execution of a final written agreement,' a strong inference is made that they 'do not intend to be bound by earlier negotiations or agreements until the final terms are settled.'"[15] Thus, care should be taken that making a preliminary agreement subject to the execution of a future definitive agreement does not automatically result in a non-binding agreement.[16]

Finally, titling an agreement 'non-binding' may not be sufficient to ensure that it will be deemed so by a reviewing court. For example, if the 'non-binding' preliminary agreement is used by one party in some way to encourage the other party to do something in reliance on the terms set forth in the preliminary agreement, a court might enforce the terms based on that detrimental reliance or some other equitable theory. Examples of other equitable theories include:

- unjust enrichment;
- misrepresentation during negotiations;
- specific promise made and relied upon during the negotiation process;
- detrimental reliance; and
- general obligations arising out of the negotiations.

Parties should always consider use of such theories as a defence against assertions that agreements are non-binding.[17]

4.3 Obligation to negotiate in good faith

Even if a preliminary agreement is non-binding, courts may impose an obligation to negotiate in good faith towards a final agreement, regardless of whether the preliminary agreement includes any such express covenant.[18] What

15 *Mass Cash Register, Inc v Comtrex Sys Corp*, 901 F Supp 404, 415 (D Mass 1995) (quoting *Rosenfield v US Trust Co*, 290 Mass 210, 216, 195 NE 323, 325 (1935)). See also *Mendel Kern, Inc v Workshop, Inc*, 400 Mass 277, 281, 508 NE 2d 853, 856 (1987) (the fact that the parties contemplated an additional document was "cogent evidence that they did not intend that the preliminary agreement, even when accepted by [the plaintiff], would constitute a binding agreement").

16 In *Nephrology & Hypertension Specialists, LLC v Fresenius Medical Care Holdings, Inc*, 2010 WL 3069758 (SD Ohio 2010), the parties had entered into a preliminary agreement that provided in part:

> *It is understood that this letter merely constitutes a statement of the mutual intentions of the parties with respect to the matters outlined herein, does not contain all matters upon which agreement must be reached in order for the proposed transactions outlined herein to be consummated and creates no binding rights in favor of any party. A binding commitment with respect to the proposed transactions will result only if definitive agreements are executed and delivered, and then, only subject to the terms and conditions contained therein.*

The court determined that where the parties had performed some of their obligations, the foregoing statement was too broad to determine what was binding and non-binding, and therefore denied the defendant's motion to dismiss.

17 See generally, Alan Schwartz and Robert E Scott, "Precontractual Liability and Preliminary Agreements" (2007), *Harvard Law Review*, Vol 120 Number 3, p662, at 668.

18 *Id* at p675. The authors write that there has been a major shift in doctrine, creating a new default rule emerging to govern cases in which the parties contemplate further negotiations. They state that the default rule starts with the presumption that preliminary agreements typically do not create binding contracts, but requires parties to "accept a mutual commitment to negotiate together in good faith in an effort to reach final agreement". The authors acknowledge that no party has the right to demand performance of the transaction if the parties cannot ultimately agree on a final contract.

does it mean to negotiate in good faith? Some authors have analogised to labour negotiations and suggested that bad faith can be evidenced by:

- failing to move from an initial position;
- presenting unreasonable proposals or last-minute proposals;
- refusing to sign a definitive agreement when all business terms have been agreed upon;
- using delay tactics; or
- sending a representative who has no authority to bind a party to negotiate.[19]

The implied duty of good faith and fair dealing is a well-established doctrine in contract law.[20] This notwithstanding, the interpretation of this doctrine can vary widely by state jurisdiction relative to commercial transactions and the parties involved. For example, reviewing courts in states such as Delaware (where many companies incorporate) will closely examine the intent of the parties to determine whether an action violates the covenant and will not apply the doctrine in contravention of an express provision to the contrary.[21] Contrast that with states such as Texas, where "there is no general duty of good faith and fair dealing in ordinary, arm's-length commercial transactions".[22] At the other end of the spectrum, Arizona (where substantial exploration and mine development occurs) recognises that a duty of good faith and fair dealing is imposed in every contract.[23] In fact, Arizona's only consistent limitation on the duty of good faith is to prevent it from "directly contradict[ing] an express contract term".[24]

Without any guidance from the parties in their writing, a reviewing court may conclude that "parties can bind themselves to a concededly incomplete agreement in the sense that they accept a mutual commitment to negotiate together in good faith in an effort to reach final agreement within the scope that has been settled in the preliminary agreement".[25] For example, the court in *Howtek, Inc v Relisys*, 958 F Supp 46 (DNH 1997) stated that:

> *The fact that an agreement grants a party some degree of discretion in performing does not render the agreement unenforceable. In such a situation, "the parties' intent to be bound by an enforceable contract raises an implied obligation of good faith to observe reasonable limits in exercising that discretion, consistent with the parties' purpose or purposes in contracting."*

---

19 See Reeves, at 2-20 to 2-22.

20 The Restatement (Second) of Contracts § 205 (1981) provides: "Every contract imposes upon each party a duty of good faith and fair dealing in its performance and its enforcement." The Uniform Commercial Code also defines 'good faith' as "honesty in fact and the observance of reasonable commercial standards of fair dealing". UCC § 1-201(b)(20)(amended 2003).

21 See *Airborn Health, Inc v Squid Soap, LP*, 984 A 2d 126, 145-46 (Del Ch 2009).

22 *Formosa Plastics Corp USA v Presidio Engineers and Contractors, Inch*, 960 SW 2d 41, 52 (Tex 1998).

23 See *Rawlings v Apodaca*, 151 Ariz 149, 163, 726 P 2d 565, 579 (Ariz 1986).

24 *Bike Fashion Corp v Kramer*, 202 Ariz 420, 46 P 3d 431, 434 (Ariz Ct App 2002); see also *Wells Fargo Bank*, 201 Ariz 474, 38 P 3d 12, 29 (Ariz 2002).

25 *Teachers*, 670 F Supp at 498.

While this court found no obligation to engage in an unremitting effort to say 'yes', it did require the parties to demonstrate a genuine willingness to come to a reasonable agreement.

Importantly, if a court finds that there is a duty to negotiate towards a final agreement in good faith, it will not require the parties to reach a final agreement.[26] In such cases, however, courts may award the innocent party expenses incurred in negotiating in reliance on the contractual assurance that the other party would negotiate in good faith.[27]

In contrast, where the duty to negotiate in good faith towards a definitive agreement is an express provision in a preliminary agreement, courts may refuse to enforce the provision in the absence of specific standards applicable to the negotiation.[28] For example, the court in *2004 McDonald Ave* stated that a clear set of guidelines was essential to measure a party's efforts before the court would enforce the obligation.[29]

From a practitioner's perspective, if the parties do not intend to have a duty to negotiate towards a definitive agreement, they should expressly state such in the preliminary agreement. The more expressly the provision is addressed, the more likely a reviewing court will not impose any such obligation as a result of the implied covenant of good faith and fair dealing.

5. How long do preliminary agreements last?

The term of a preliminary agreement depends on the intent of the parties and should be clearly set forth therein. If a preliminary agreement is non-binding and allows for termination at any time, problems may arise if negotiations stall or changed circumstances become apparent.

Commonly, parties will specify that a preliminary agreement will remain in effect until a definitive agreement has been negotiated, but ignore the consequences of failure to reach a definitive agreement. If no definitive agreement is executed and the preliminary agreement contains an exclusivity or standstill clause, the agreement should specify what happens following the expiration of those timeframes. Does this mean that the preliminary agreement has terminated as to all provisions after the expiration of certain timeframes?

---

26 *Venture Associates Corp v Zenith Data Systems Corp*, 96 F 3d 275 (7th Cir 1996).

27 See Reeves, at 2-23, citing *Continental Business Enterprises, Inc v United States*, 452 F 2d 1016 (Ct Cl 1971) (the submitter of a proposal for a negotiated government contract may recover the expense of preparing a proposal if the government violates its statutory duty to negotiate).

28 See *2004 McDonald Ave Realty, LLC v 2004 McDonald Ave Corp*, 50 AD 3d 1021, 858 NYS 2d 203 (2d Dep't 2008).

29 See also *A/S Apothekernes Laboratorium for Specialpraeparater v IMC Chem Grp, Inc*, 873 F 2d 155, 158–59 (7th Cir 1989) ("[i]n the absence of any agreed upon terms or even a general framework within which to conduct the negotiations, the parties were free to insist on or reject any proposed terms to the contract that they wished")); and *Clark Resources, Inc v Verizon Business Network Services, Inc*, 2011 WL 1627074 (MD Pa 2011) (the court refused to enforce a covenant to negotiate in good faith without some "framework" or "objective criteria" by which a court can measure whether the agreement to negotiate in good faith has been breached).

Are there some other provisions that survive? Again, careful drafting is important.

Finally, if the parties execute a definitive agreement, there will often be a 'merger' or 'integration' clause, which provides that the definitive agreement contains the entire agreement of the parties and supersedes all other agreements on the same subject. Such clauses, while of a boilerplate nature, may not suit the parties' deal structure; therefore, consideration must be given to the provisions unique to the preliminary agreement that remain enforceable after execution of a definitive agreement.

6. Conclusion

Preliminary agreements can be helpful in consummating a transaction or mineral exploration, development, finance, acquisition or divestiture. They are best used with forethought and planning relative to which provisions will be binding on either party so that there are no unintended consequences. The first question to consider is whether the parties should enter into a preliminary agreement and if so, why. If there is suitable justification to enter into a preliminary agreement, the next question to consider is what provisions, if any, should be binding on the parties. Finally, care should be taken to expressly terminate any binding obligations upon execution of a definitive agreement or otherwise specify survival of any carry-over terms.

Option agreements in the mining industry

Shea Small
Roger Taplin
Adam D Wanke
McCarthy Tétrault

1. Introduction

Mark Twain once wrote, "Never allow someone to be your priority while allowing yourself to be their option." Twain's advice may have force in one's personal affairs, but it is less germane when a mineral project owner interacts with a party that seeks an interest in the owner's project. The project owner (or optionor) believes in the value of its property, considering it a prize which just requires some financial and/or technical input to take it to the next level. The investor (or optionee) is interested in the potential of the project, but does not want to overcommit itself, preferring to test the waters before investing significant funds. These differing objectives of the optionor and optionee provide the imperatives for an option agreement transaction.

An option agreement – sometimes referred to as an earn-in agreement or farm-in agreement – is probably one of the most common forms of agreement in the mining industry. It is an agreement between the optionor and the optionee which allows the optionee to earn some type of interest in the optionor's property in exchange for certain benefits granted to the optionor. Option agreements tend to have four primary elements and a collection of less critical but still important features:

- Interest: The optionee will need to understand the nature of the mineral right that is being acquired. What is the nature of the optionor's ownership of the mineral project? Are there any encumbrances against the property that need to be provided for? The parties will also need to negotiate the level of interest that the optionee is acquiring in the mineral project – is it a straight-out 100% acquisition or is it a smaller percentage interest through a joint venture to be established afterwards? Is the interest being acquired all at once or is the option broken up into mandatory and optional ownership step-up stages?
- Consideration: The type of benefit that the optionor will receive as part of the exercise of the option by the optionee can include direct cash payments or issuance of securities of the optionee, commitments by the

optionee to perform certain work on the mineral project or the delivery of studies on the mineral project.

- Operator: Who will conduct work on the property during the course of the option term? The optionor likely has more familiarity with the mineral project and the region, and often has all of the inherent authorisations to conduct work on the mineral project by virtue of its ownership. The optionee will quite often be a larger international 'major' in the mining industry, with highly specialised and experienced mining staff and expansive internal policies.
- Project: Who will hold title to the property during the option term? As per the theme of the introductory paragraph, the mineral project is probably of central priority to the optionor and it may be anathema for the optionor to transfer title in its mineral project to the optionee before the optionee has even begun work. At the same time, the optionee will want some guarantee that it is not spending its money in vain and that there will be an interest in a project at the end of the option period. From the optionee's perspective, the best way to guard against events such as the insolvency of the optionor is to hold the title to the mineral project in its name, subject to an obligation to return the mineral project to the optionor if it decides not to exercise the option.
- Other: Quite often, the parties to a negotiated option agreement will pop the cork and shake hands once the issues above are settled and regard the rest of the agreement as 'just boilerplate'. However, that is unwise. How is the option exercised? Is there an area of interest? What are the eligible expenditures and how are they confirmed? What happens if the option terminates before exercise? What happens if there is a blockade? These and other matters should also be considered with care by the parties.

2. Common features of an option agreement

2.1 Option interest – what's your interest?

One of the most important commercial terms for any option agreement is the level of interest that the optionee will receive if it satisfies all of the requirements to exercise the option. The earned interest can range anywhere from a small minority interest up to a full 100% interest over the property. In the case of a 100% acquisition, things are a lot simpler, since the option is essentially a delayed purchase transaction and the parties do not need to regulate their post-option exercise relationship.

If an optionee is acquiring less than a 100% interest, once the option is exercised, the parties will be in a joint venture and will need to determine how the joint venture is governed. While the form of joint venture to be entered into and the terms of the joint venture agreement are beyond the scope of this chapter, one important aspect that needs to be addressed is how the joint venture terms are described in the option agreement. There are essentially three general approaches to setting out the terms of the joint venture in the option agreement:

- Stay silent on the terms and simply insert an obligation on the parties to negotiate the joint venture form and terms following exercise of the option;
- Include a near-complete set of joint venture documents that fully captures the nature of the parties' relationship as joint venturers; or
- Attach a detailed term sheet that captures the essential elements of the joint venture, such as initial interests, form of joint venture, governance and management structure, funding requirements and dispute resolution.

The first approach involves the least upfront cost and time commitment, since the parties are focusing on the option agreement terms; theoretically, this approach also has the advantage of letting the parties negotiate a joint venture agreement at a time when they have a much better idea of the size, scope and nature of the mineral project. However, putting off until tomorrow what can be done today has its disadvantages – the most significant of which is that, based on common law principles, once the option is exercised, the agreement to negotiate a joint venture agreement is an 'agreement to agree' and is most likely not binding. While it can be argued that the parties did intend to create some form of binding contractual joint venture relationship, none of the essential terms will have been agreed; nor will there be sufficient certainty as to the content of such an agreement. Further, should the optionee's work on the mineral project reveal a project that has good potential, any initial cost and time savings may evaporate as an optionor, having confirmed that it has a good project on its hands, may be much less motivated to reach a joint venture agreement. Similarly, the optionor may find that the optionee is much less motivated to give the optionor a seat at the table in terms of managing the joint venture and, assuming that the optionor is now the minority party in the joint venture, may be faced with an optionee which resists all but the most basic minority protections. Our view is that it will be a very rare case where the significant disadvantages of leaving out the joint venture terms outweigh the benefits of signing a completely standalone option agreement.

The greatest advantage of the second approach is that the parties can be certain as to how their eventual joint venture will function and what each of their respective rights and obligations will be. From a lawyer's perspective, this is the ideal approach, given that there is no contractual gap between what is agreed at the option stage and what is agreed for the joint venture stage. However, there are two significant drawbacks to this approach: time and money. Spending weeks and incurring legal fees negotiating a joint venture agreement for something that is, as yet, far from certain may seem a less than ideal use of the parties' resources. Junior companies which are acting in the role of the optionor will be especially reluctant to spend time and money negotiating a full back-end joint venture agreement when the international major optionee can cut its losses and abandon the option at any time. However, a mistake that companies often make in trying to have their joint venture cake and eat it is to drop in a standard form joint venture agreement that one of the parties already has as a precedent or that is commonly known within the industry, such as the Rocky Mountain Mineral Law Foundation form agreements. While such internal precedents or industry organisation forms are valuable tools and provide a good starting point, they are not usually 'one size fits all' and it can be very dangerous to bind the parties to govern their joint venture in this way if adequate thought has not been given to how these precedents address the particular circumstances of the project in hand and the circumstances of the parties involved. However, we have seen a variance on this approach where the option phase is built into and is part and parcel of the joint venture agreement – essentially, the parties execute the full joint venture agreement at the outset, but the optionee's interest is subject to revocation and the joint venture agreement to termination if the optionee terminates the agreement or defaults on exercise of the option. Thereafter, the parties function similarly to how they would in the case of an option agreement – the optionor will need to have its portion of joint venture costs funded by the optionee until a 'joint funding date' is reached, where the optionee has spent a pre-determined amount of money or satisfied other criteria. Should the optionee fail to do so, the joint venture is dissolved and the optionor's interest in the mineral properties is restored. This 'earn-in' joint venture approach may be particularly suited to option/joint venture deals that involve incorporated offshore joint venture companies with an onshore project owner, since the parties can create the appropriate structure from the outset and not have to worry about scrambling to create all the proper corporate or structural buckets when the option is eventually exercised.

The third approach is something of a 'Goldilocks' approach, given that it avoids the uncertainty pitfalls of the first approach and does not require the same amount of time and legal costs as preparing a full-blown joint venture agreement under the second. Notwithstanding the attractiveness of this approach, the joint venture term sheet should be sufficiently certain and detailed and cover all important structural, governance and funding processes that will form the core of the joint venture agreement. This will make the joint venture agreement negotiations that much easier once the option is exercised.

2.2 Work and funding requirements – you've earned it

The parties next have to agree on the conditions that the optionee must satisfy in order to exercise the option. While each option agreement will contain different option exercise criteria, the most common earn-in requirements are a combination of the following:

- cash payments or other forms of payment (eg, equity issued by a public company) to the optionor;
- funding of work to be performed on the mineral properties; and/or
- preparation and delivery of specific deliverables, such as studies and reports on the mineral project, by the optionee.

Cash payments are fairly self-explanatory – the optionee needs to pay the optionor a certain amount of money to keep the option in good standing. From an optionee's perspective, this amount should be as minimal as possible, since it represents money spent without giving the optionee any better sense of whether the mineral project justifies further expenditure. However, even from the optionor's perspective, it would be more advantageous that the optionee 'put money into the ground' rather than make cash payments to the optionor, since the whole point of the option is to secure the funding and technical expertise of the optionee to advance the optionor's understanding of the mineral project.

By far the biggest chunk of the earn-in expenditures will most often be in the form of funding of work on the mineral properties to be completed over the course of the option period. Accordingly, it is very important that the parties spend the time and effort to understand exactly what qualifies as 'expenditures' incurred by the optionee in respect of work on the mineral project. From the optionor's perspective, it will want as much of the work expenditures as possible to relate to actual work: drilling, aerial surveys, core analysis and so on. The optionee, on the other hand, will aim to have as broad a definition of acceptable expenditures as possible, so that anything related to the work on the property is credited against the optionee's expenditure obligations. For that reason, it is our view that the parties should spend the time and effort to properly define what are and what are not acceptable expenditures, and not rely on shorthand such as "costs customarily incurred in exploration activities in the mining industry". While the latter approach seems simple and attractive, both parties will rue the day they decided to cut corners when the optionor disputes a material charge or class of charges that the optionee is claiming. Some important questions that will need to be answered include the following:

- How are related-party costs treated?
- Is the optionee permitted to claim a portion of its overhead costs as option expenditures?
- Will the optionee be permitted to charge an operator's or management fee in respect of the work completed?
- Are there statutory requirements for qualifying work commitments that will need to be met?

In cases where a substantial presence of the optionee's staff or contractors is required on site, it is fairly well accepted that the optionee will be permitted to deduct a certain amount for overhead and office expenses from its earn-in expenditures. Quite often, overhead will be captured by way of a percentage of overall expenditures actually incurred by the optionee in carrying out qualifying work that is then deducted from the optionee's earn-in expenditure obligations.

The last type of 'consideration' commonly used to exercise an option is the delivery of some kind of study or report. This can include things such as a 'scoping study', 'pre-feasibility study', 'feasibility study', 'positive feasibility study' and 'bankable feasibility study'. The most important point to be made here is not to rely on industry understanding of what any such deliverable is meant to be, but to precisely and accurately describe what is meant – preferably using the standards set out in National Instrument 43-101 in Canada, the Australasian Code for Reporting of Exploration Results, Mineral Resources and Ore Reserves or another equivalent technical standard. The optionee should strongly resist any use of the terms 'positive', since these terms connote outcomes of the optionee's work on the property that are outside of its control.

In terms of the timing of expenditures, the optionor will want to ensure that the optionee is diligently spending money on the property and conducting the work necessary to exercise the option. Most option agreements will not only have an overall aggregate funding or work commitment obligation, but also include annual funding or expenditure obligations, and the optionee will be required to satisfy both in order to exercise the option. On the flipside, if an optionee determines, after a year or two of work, that the property is a winner, it should have the ability to accelerate the option by either paying in cash the full amounts remaining outstanding to the optionor (or into the joint venture to be formed upon exercise) or ramping up operations. The last thing that either party would want is to be stuck in a pantomime trudge to the finish when the mineral project merits the earlier formation of a joint venture and ramping up towards development.

2.3 Option exercise – don't get exercised

It is important to make sure that both parties understand what actually happens once the optionee has fulfilled its earn-in obligations. There are two distinct approaches as to how to exercise an option:

- Exercise requires an 'exercise notice' from the optionee to be delivered to the optionor; or
- The option is deemed to be automatically exercised once the option requirements have been met.

From the optionor's perspective, the exercise mechanic is of less concern, since in either case whether the optionee has actually satisfied the option obligations is something that the optionor can challenge if it is of the view that such obligations have not in fact been satisfied. The optionee may at first blush prefer the automatic exercise option for its simplicity, given that the notice approach requires extra steps and that diligence be paid to meeting notice delivery timeframes and formalities. While those are valid concerns, our view is that an optionee would be better served by having the final exercise step be within its control. It may well be that while the optionee has satisfied the work commitment obligations necessary for an option to be exercised, it has discovered some legacy environmental liability that is sufficiently material to deter it from wanting to become an owner of an interest in the mineral project – in the case of an automatic exercise, the optionee may be unable to avoid becoming an owner.

2.4 Areas of interest – all you see before you is yours

Some option agreements will have an 'area of interest' provision, whereby any acquisitions of an interest within the defined area of interest will cause the acquired property to become part of the optioned property. An optionor may want to include an area of interest provision to protect itself from the optionee (which likely has more financial resources at hand) snapping up claims surrounding the option property once it becomes clear that the mineral project has merit. At the same time, the optionee may wish to define an area of interest if it anticipates that an eventual project may have a wide infrastructure footprint outside of the current optioned property. In either case, the parties should ensure that:

- the process for the exercise of any area of interest is clearly laid out;
- notice requirements are specific; and
- the election as to whether to include the subject property within the option agreements is made solely at the discretion of the non-acquiring party.

A detailed examination of area of interest clauses is beyond the scope of this chapter and readers are strongly advised to exercise caution with these provisions, which have been a fruitful area for litigators.

2.5 Indemnification – someone's got my back

Something that is often overlooked in option agreements is indemnification of the optionor in respect of the conduct of work by the optionee over the course of the option period. As environmental regulators constantly seek to extend the net around who can be found liable for environmental damage that occurs as a result of any mining related activity on a property, the optionor should not be lulled into a false sense of security by thinking that just because it is not directly doing the work (or not directly holding the legal title to a particular property, depending on how the option transaction is structured), it does not need additional protection in the form of indemnification by the optionee in the option agreement. The optionor will want to ensure that the indemnity is broad and captures any liability arising from the work conducted by or at the direction of the optionee, or from any entry onto the property by the optionee or its agents. By the same token, the optionee will often seek a similar indemnity for all work conducted on the property prior to the effective period of the option agreement, and for any work conducted by any person should the option be terminated without being exercised.

In the event that it is discovered that some environmental liability has arisen on the property, it may be difficult to establish when and by whom the underlying damage was caused. Where potentially significant environmental liabilities are implicated, a bright line can be established in advance by having the optionee conduct an environmental baseline study that describes the condition of the option property prior to the option work thereon commencing – a similar study may be conducted again should the option be terminated. While conducting these studies may be cumbersome, it is in the interests of both parties to clearly delineate what their potential obligations may be in respect of environmental liability; the parties can agree to have the costs thereof deducted from the optionee's earn-in obligations or otherwise shared among them.

2.6 Termination – it's not you, it's me

While hope springs eternal when the parties first sign an option agreement, the fact is that in most cases, the optionee will determine that the results of the option work do not satisfy its internal requirements for continuing with the project. A well-drafted option agreement will contain a specific process with defined timeframes as to what happens next. There will be a tension between the optionee's desire to have as little ongoing liability as possible as a result of the termination and the optionor's desire to minimise its exposure to ongoing costs. An optionor may seek to negotiate a minimum funding or expenditure guarantee – if the optionee terminates the option early without having met this minimum target threshold, it will be responsible for paying the difference to the optionor. The optionor may also seek to ensure that the optionee has paid up all minimum statutory contributions to keep the mineral tenure in good standing and has satisfied minimum statutory work commitments for a set period after termination of the option (usually not more than 90 days), and that the optionee removes its equipment and other structures from the option property and leaves the property in a safe condition and in compliance with environmental law in respect of any work carried out thereon by the optionee.

Other matters – such as whether the optionee can retain the information generated as a result of its operations on the mineral project, how long an area of interest survives following termination and obligations with respect to any core samples in the possession of the optionee – should also be clearly addressed in the termination provisions.

3. A closer look at the operator and title

3.1 Operatorship – smooth operator

One of the most significant decisions to be made when the option agreement is negotiated is whether the optionor or the optionee will act in the role of operator of the project during the option period. While the optionor, which usually knows the property very well, may wish to retain direct control over the work carried out on the property, in some cases the best long-term choice may in fact be conceding the role of operator to the optionee. In most cases the optionee will be a major mining company with the operational expertise and mining and technical knowledge that are likely to have been one of the major attractions for the optionor in entering into the option agreement in the first place. Also, where exercise of the option requires the performance of work or the expenditure of money, the optionor will generally want control of those activities.

If the optionor nevertheless acts as the operator under the option agreement, the optionee will want to seek clear and detailed provisions dealing with oversight, reporting and accounting matters, given that all that the optionee will effectively be doing during the option period is feeding money to the property owner and hoping that something comes out on the other side that justifies the expenditure. The optionee will want guarantees (or a right to step into the operator role should the optionor fail to satisfy certain minimum requirements) from the optionor that it will conduct the amount and quality of work necessary for the optionee to be able to satisfy its funding obligations; otherwise, the optionee could be faced with the extraordinary situation where it fails to meet an annual expenditure requirement necessary to maintain its option under the terms of the option agreement simply because the optionor did not do enough work. In such circumstances, the optionor may also demand an extension to any applicable option timelines.

By the same token, if the optionee is the operator, the optionor too will want to ensure that the optionee is not given free rein over the property and that specific standards of performance are imposed, together with a reporting and accounting regime that allows the optionor to keep an eye on the optionee's performance. At the same time, the optionee should take care that the option agreement allows it not only the operational freedom necessary to satisfy its earn-in obligations, but also the flexibility necessary to adjust and reposition operations as the optionee gets a better sense of the four corners of the project. The optionee should also try to avoid absolute obligations to achieve specific technical milestones (eg, an obligation to drill a hole to a specific depth), because this may prove to be an inefficient use of resources once the optionee better understands the property; instead, specific technical achievements that the optionee is required to achieve should be framed on a 'reasonable efforts' basis. It may be trite, but if the optionee is the operator under the option agreement, the optionee should also ensure that the option agreement expressly provides that it has the right to enter the property, to have quiet possession thereof and to conduct work thereon.

Regardless of whether the optionor or the optionee ends up acting as the operator during the option period, the non-operator will need to ensure that various protections are inserted into the option agreement to safeguard its interest in the mineral project, to confirm that operations are proceeding as expected and to make sure that money is being spent in a way that actually advances the project. Basic obligations that the non-operator will want to impose on the operator include:

- obligations to keep the mineral tenure in good standing and free and clear of any encumbrances other than permitted encumbrances specifically defined in the agreement;
- covenants not to engage in any activities that might jeopardise the mineral tenure;
- covenants to maintain adequate insurance;
- covenants to conduct all operations in accordance with applicable law and good exploration and mining practice; and
- reporting obligations to ensure that each quarter, or some other such similar period, the optionee delivers to the optionor detailed progress reports on operations completed to date, as well as more detailed annual reports that outline completed operations, as well as describing operations forecasted to be completed in the following year. The non-operator may also seek to get copies of all raw data obtained during the course of the operator's operations. In either case, the non-operator should have appropriate audit and inspection rights.

While the sky can be the limit in terms of robust reporting and oversight regimes, most non-operators will likely resist extreme oversight and reporting regimes. Each side should seek to avoid a reporting regime that is overly onerous – frequent detailed reporting will consume a lot of time and money, both of which would be better spent by the operator putting drills into the ground.

3.2 Property title – what's in a title?

While also a binary choice, the question of who will hold title to the mineral properties during the option period is sometimes more difficult to settle than the question of who will act as operator. Most optionees which are international majors will be wary of spending large amounts of money on a project in which they do not hold legal title to the mineral properties. The optionee will have two significant fears:

- The optionor may become insolvent during the option period; and
- The optionor will attempt to avoid transferring the property to the optionee or to the joint venture entity once the option is exercised.

As a result, the optionee will often argue that the legal title to the mineral properties be transferred to it to hold, subject to an obligation to return the mineral properties to the optionor if it does not exercise the option. Under such a scenario, if the optionee ultimately exercises the option, it will:

- retain the legal and beneficial title to the mineral properties in the case of a 100% option;
- hold the legal title to the mineral properties in trust for the joint venture partners in the case of an unincorporated joint venture; or
- transfer title to a new entity that is formed to hold the mineral properties as part of the newly formed joint venture between, and for the benefit of, the optionor and the optionee.

Either way, it is protected.

The option of holding the legal title in the property 'in trust' for the benefit of both parties is the simplest mechanism by which the optionee can take ownership of the mineral properties under an option agreement; but this mechanism will not be available in jurisdictions that do not recognise the concept of a split between legal and beneficial ownership, such as in civil law jurisdictions. In such jurisdictions, a solution will require more complex hoop-jumping. In some cases, the party that does not hold title can take a security interest in the mineral properties or in the entity that holds the mineral properties. In other jurisdictions, a registration can be made against the title to the mineral properties that recognises the other party's interest in the mineral properties and protects such interests against other third parties. In yet other circumstances, the party that holds the title to the mineral properties will sign and provide to the other party transfer documents that grant the right to the other party to effect a transfer of the mineral properties by 'registering' the signed transfer documents if certain events of default occur as defined in the option agreement. Whatever the actual steps taken to recognise the interest of the party that will not hold title to the property, such party should seek the advice of local counsel in the jurisdiction where the property is located to determine what action will best protect its interest in the property.

While the optionee may take comfort in completing some or all of the steps described above that are available to it in a particular jurisdiction, this may prove to be cold comfort if, in the event that the optionor is insolvent, a bankruptcy court determines that it can set aside or ignore any claims of interest by the optionee and treat the property as part of the assets of the bankrupt optionor. This risk becomes more marked the further into the option period the parties have progressed, since the work funded on the property by the optionee will likely have crystallised the ultimate value of any mine to be built on the property, thus making it a more valuable asset to the bankrupt estate. Depending on the jurisdiction, it may be the case that even if the optionee holds the mineral title in trust for both parties, such an arrangement is not ironclad protection against an ambitious liquidator that is seeking to increase the pot of available assets to divide among clamouring creditors. The impact of the insolvency of the optionor on a property subject to an option agreement is a complex topic and one that is heavily dependent on the particular practices of the host jurisdiction. The optionee should thus invest time and money to get a view from local counsel on the best way to manage such risk.

As discussed earlier in the context of the process for the exercise of the option, whatever the final outcome with respect to where title to the mineral properties will rest, both parties will want to make sure that the timing for when the property is transferred to the optionee or the joint venture entity is clearly set out in the option agreement.

4. Conclusion

We began with Mark Twain and will end with him. It is a source of debate whether he ever really did define a 'gold mine' as "a hole in the ground with a liar on top"; but whether or not he did, the sentiment is apposite in the context of option agreements. A well-thought-out agreement will help to keep all parties honest and may help to result in a lucrative hole in the ground.

Joint ventures

Stuart MacGregor
Marie-Christine Moon
Clayton Utz

1. Introduction

This chapter explores the differences between the main agreements used to document joint venture projects in the mining industry and the key provisions during the project's lifecycle.

The most common types of mining joint ventures are unincorporated. This chapter examines some of the key elements of joint venture agreements, including the mechanisms used to confer on the joint venture parties their respective interests in the project, as well as the mechanisms used to control and implement the project, including the appointment of a manager, the management committee and programmes, budgets, cash calls and authorities for expenditure (AFEs).

Three contentious elements of joint venture agreements are then examined. The first is assignment and pre-emptive rights. It is usually difficult to get joint venture parties to think at the start of their relationship about what will happen when one party wants to end the relationship; but the structure and drafting of a pre-emptive rights clause must be carefully considered and documented to provide a clear and unambiguous process that will withstand close scrutiny – and hopefully avoid disputes. Experience shows that while this is generally one of the more heavily negotiated clauses of a joint venture agreement, it is not until a party actually seeks to dispose of its interest that the clause is put to the test.

The second is the default of a joint venture party. The structure and drafting of the default clause must be carefully considered, especially where a defaulting party may be required to assign its interest to the non-defaulting parties for a discounted price – or no price at all. The intent behind the clause and the particular circumstances of the joint venture will be factors in assessing the clause within the context of the legal principles of penalties and relief against forfeiture. This chapter touches on these legal principles and highlights the importance of striking the right balance when negotiating and drafting default mechanisms in resource joint venture projects.

The third key issue is the post-production phase of the mining joint venture and issues for consideration in relation to long-term rehabilitation, monitoring and eventual hand back of the project to the local regulatory authority.

1.1 What is a joint venture?

Two key types of joint ventures are used in the mining industry: the unincorporated joint venture and the incorporated joint venture. An unincorporated joint venture is a contractual arrangement between the participants for the exploitation of separately owned interests in a resource; whereas an incorporated joint venture involves the formation of a separate legal entity to own and exploit the resource.

The decision on whether to set up an incorporated rather than an unincorporated joint venture is often driven by local jurisdictional law concerns such as tax and government restrictions on certain types of investment structures.

The reason why unincorporated joint ventures are preferred in the mining industry is for their unique ability to structure collaborative projects in a manner that permits a high degree of independence between participants, while allowing them to share their commercial resources. The joint venture model addresses many unique challenges posed by resources projects: their speculative nature, expense and high risk, as well as the technical demands of the industries in the sector. By adopting the joint venture model, parties may cooperate in a discrete project without entering into a continuing commercial relationship. Parties enjoy a high degree of contractual flexibility in addressing the specific demands of a given project and, through the joint venture, share costs, risks, technical expertise and assets – including the opportunity to exploit a mining tenement owned by another party.

The term 'joint venture' is commonly used; but in most jurisdictions, it is not a clear, legally defined concept and is not necessarily correctly or consistently understood. In some jurisdictions, joint ventures are covered by partnership laws (albeit having some different characteristics from a partnership); while in other jurisdictions, such as Australia, the unincorporated joint venture agreement will often contractually deny the status of a partnership and seek to avoid the attribution to the participants of the concepts of fiduciary duties, agency and joint liability.

The distinction between a partnership proper and a joint venture was recorded in Scottish law back in the 1760s, in JE Erskine's *An Institute of the Laws of Scotland* (new ed 1871). The Scottish concept of 'joint adventure' or 'joint trade' closely resembles the structure of the modern Australian mining joint venture in many respects.[1]

In an Australian context, 'joint venture' can best be described as a term that:

connotes an association of persons for the purpose of a particular trading, commercial, mining or other financial undertaking or endeavour with a view to

1 Gerald L J Ryan, "Joint Venture Agreements" (1982) 4(1), *Australian Mining and Petroleum Law Journal* 101, 106.

mutual profit, with each participant usually (but not necessarily) contributing money, property or skill ... The borderline between what can properly be described as a "joint venture" and what should more properly be seen as no more than a simple contractual relationship may on occasion be blurred.[2]

In the United States, the concept of a joint venture is not clearly defined. Although it is closely associated with the laws of partnership,[3] since the 19th century the United States has developed a body of case law that distinguishes joint ventures from partnerships.[4] However, the foundations of a joint venture are clearly based on the laws of partnership and extend to fiduciary relationships.[5]

In Canada, the definition of a 'joint venture' is likewise unclear and the Canadian courts have adopted a similar approach to that taken in the United States.[6]

The term 'joint venture' is used throughout the world to loosely describe a variety of other relationships between associated parties, including partnerships, trusts and strategic alliances.

Mining joint ventures usually have two main features that distinguish them from other relationships:

- A joint venture is not usually the principal commercial activity of each joint venture party. Instead, it is an extension of each party's commercial dealings where the resources, identity or skills of each participant complement those of the other participants. Taken together, there is some commercial or strategic value to justify the alliance.
- The undertaking that is the subject of the joint venture is usually a single business project of a limited duration or a business undertaking of a limited scope, rather than a series of projects.

It is important to distinguish the joint venture itself from the underlying entity or entities (the participants) that carry out the joint venture.

2. Why choose a joint venture?

There are several reasons why resource companies choose to associate themselves in a joint venture structure. A joint venture structure allows the participants to:

- spread exploration risks;

---

2 *United Dominions Corporation Ltd v Brian Pty Ltd* (1985) 157 CLR 1, 0 (Mason, Brennan and Deane JJ, with whom Gibbs CJ agreed).

3 *Firestone v VanHolt*, 186 SW 3d 319 (Mo Ct App WD 2005).

4 Maree C Chetwin, "Joint Ventures – A Branch of Partnership Law" (1991) 16(2), *University of Queensland Law Journal*, 256, 257.

5 J Taubman, "What Constitutes a Joint Venture?" (1956 41 Cornell LQ 641).

6 Melanie A Shishler, "The Graham Decision Revisited: The Fading Promise Of The Joint Venture As A Distinct Legal Concept" (1998) 31 *Canadian Business Law Journal*, 118, 118; *Central Mortgage & Housing Corp v Graham* (1973) 43 DLR (3rd) 686.

- spread high capital and operating costs;
- access greater financial resources than a single party could access;
- achieve economies of scale and take advantage of size;
- gain access to new customers;
- create stronger competitive units;
- create synergies;
- transfer technology and skills; and
- diversify into new areas.

3. Joint venture structure

3.1 Incorporated versus unincorporated

Parties wishing to associate as joint venturers should make sure that they understand the legal implications of the structure, and that the commercial arrangements they enter into do not inadvertently lead to unintended tax consequences or other legal obligations.

Joint ventures can be incorporated or unincorporated. However, the most common structure used by participants in the mining industries, which is the focus of this chapter, is the unincorporated joint venture.

3.2 Incorporated joint venture

An incorporated joint venture is a special purpose company owned by the participants (shareholders) in the joint venture. It bears the characteristics of any other incorporated entity. The joint venture company will pursue the project activities in its own right and it, rather than the participants, will generally be the selling party under any product sales or off-take agreements. A board of directors administers the business and a company constitution will set out powers to appoint directors, restrictions on capital raising and pre-emptive rights restricting share transfers. The joint venturers, as shareholders in the company, will receive dividends from the profits of the company derived from the sale of the product, rather than the product itself or the immediate proceeds of sale of the product. Typically, a shareholders' agreement will govern the business and management objectives of the joint venture and set out the dividend policy and joint venture manager's responsibilities. Incorporated joint ventures are less common in the mining industries.

3.3 Unincorporated joint venture

An unincorporated joint venture is a contractual relationship and nothing more. While unincorporated joint ventures are commonly mistaken as separate legal entities, they are merely a contractual arrangement between two or more participants which associate themselves for the purpose of carrying out a particular venture or activity, with each participant contributing funding, property or skill. The terms of the contractual relationship are set out in a joint venture agreement.

As an unincorporated joint venture is a creature of contract, the liability of the participants needs to be set out in the joint venture agreement. In jurisdictions such as Australia, unlike partnership agreements – where partners, as a matter of law, are jointly and severally liable for the expenses and liabilities of the partnership – joint venture agreements typically provide that the participants will be severally liable for all expenses and for any liabilities that arise in relation to the joint venture in accordance with their respective interests in the joint venture. The partnership structure is generally avoided in resources projects because of the resulting joint and several liability, the fiduciary duties that arise between commercial partners and the tax treatment of partnerships.

Fiduciary obligations can and do exist in unincorporated joint ventures, depending on the rights and obligations of the participants set out in the joint venture agreement. It is very common in unincorporated joint venture agreements for the participants to include wording to negate the existence of fiduciary obligations. However, even with such wording, fiduciary obligations may still exist.

Another important characteristic of an unincorporated joint venture is that the participants have the right and obligation to take and separately dispose of their percentage share of the product of the joint venture, rather than a share of the profits of the joint venture. This is an important distinction: if participants share profit, it is likely that the relationship will be a partnership which, as mentioned above, can have unintended consequences, such as joint and several liability, fiduciary duties and various tax consequences.

Unlike an agency arrangement, the manager of the joint venture and the participants in the joint venture do not automatically have the authority to bind each other, unless the joint venture agreement provides otherwise. In resource joint ventures, it is common for the manager of the joint venture to have authority to enter into contracts for joint venture purposes as agent for and on behalf of the joint venture participants, subject to agreed parameters set out in the joint venture agreement.

The property of an unincorporated joint venture will be held in accordance with the terms of the joint venture agreement. Typically, in resource joint ventures, either property is owned by the participants as tenants in common, in accordance with their respective interest in the joint venture, or the manager of the joint venture will hold property in its name, but on trust for the participants in the joint venture.

Importantly, in a number of key mining jurisdictions, an unincorporated joint venture is not recognised for tax purposes. Instead, the participants are taxed at their own level and not as a joint venture. Deductions may be available for certain expenditure and there is the potential to offset tax losses against other assessable income, particularly in the early stages. Each participant may choose separate financing and accounting methods that are best suited to its particular circumstances.

3.4 Summing up

As shown in Figure 1, the participants in an incorporated joint venture incorporate a joint venture company to undertake joint venture activities and cooperation is governed by a shareholders' agreement. The ownership interest is determined according to the number of shares that each party holds in the joint venture company, which is itself a separate legal entity.

However, the parties to an unincorporated joint venture do not incorporate a separate joint venture company. Instead, they enter into a joint venture agreement which contractually sets out their rights and obligations, and appoint a manager to carry out joint venture activities. The ownership of the joint venture assets is retained by the participants and the joint venture has no separate legal identity.

Figure 1. Incorporated and unincorporated joint venture structures

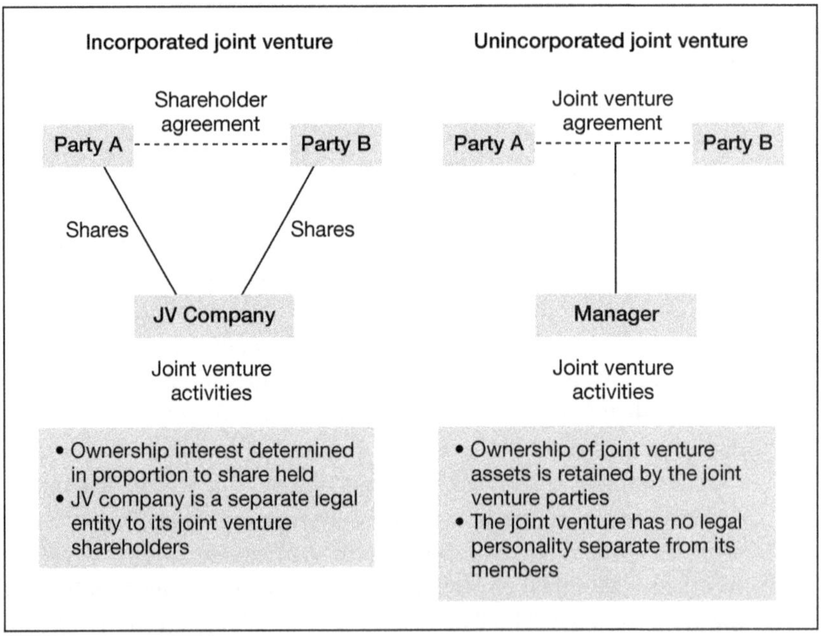

4. Joint venture agreements

In general, the agreement documenting a mining joint venture is referred to as a 'joint venture agreement'. A joint venture agreement governs the joint venture relationship between the participants and establishes the purpose and scope of the joint venture. There is no such thing as a standard joint venture agreement: every document is made to order. At the very least, a well-drafted joint venture agreement should set out:

- the scope, purpose and duration of the joint venture;
- the equity interests of each participant;
- the assets belonging to the joint venture;
- how the assets of the joint venture are owned;
- the operation, management and control of the joint venture, and the liability of the manager (although this can be in a separate management agreement);
- the entitlement of each participant to the 'product' of the joint venture;
- the process for expenditure of funds;
- the apportionment of liability;
- the consequences of default;
- how the participants in the joint venture can assign or transfer their joint venture interests;
- any restrictions on participants withdrawing from the joint venture relationship; and
- how the joint venture relationship will come to an end.

A management agreement is an agreement where the participants in the joint venture appoint someone to operate and manage the day-to-day affairs of the joint venture. A separate management agreement will be required only if the manager is not a participant in the joint venture; otherwise, the management provisions can be included in the joint venture agreement.

There follows a discussion of some of the most important considerations when negotiating an unincorporated joint venture agreement for a mining undertaking.

4.1 Joint venture interest

A joint venture agreement must clearly identify the participants in the joint venture and must set out the equity interest of each participant in the joint venture, which is often referred to as a 'participating interest' or 'percentage interest'.

As one of the main reasons for associating as unincorporated joint venturers is to allow participants to limit their risk exposure to their respective participating interests in the joint venture, the joint venture agreement should also provide that each participant's liability is several in the proportion of its participating interest in the joint venture. If percentage interests are not expressly specified, legally the participants will be jointly and severally liable, as it is not possible to determine their respective liabilities.

4.2 Scope of the joint venture

The joint venture agreement must clearly set out the scope of the joint venture – that is, what it is that the joint venture participants are associating themselves to do. As an example, the participants might associate themselves as unincorporated joint venturers to pursue the following objectives, subject to the terms of the joint venture agreement:

- to explore for product in, and appraise, the licence area;
- to develop and produce any reserves in the licence area;
- to acquire, construct, install or develop any facilities necessary or desirable to implement the above objectives; and
- to undertake any related activities that the management committee unanimously decides to pursue.

To avoid the joint venture being a partnership, it is important that the scope of the joint venture does not extend to the joint marketing of the product of the joint venture. Each participant must separately and severally sell its own share of product (although it is common for each participant to appoint a common marketing agent for this purpose).

4.3 Manager

When forming a joint venture, participants must decide whether one of the participants will be appointed to manage the joint venture or whether a third party will be appointed. In general, a mining joint venture will appoint a 'manager'. A separate management agreement will be required if the manager is not also a participant in the joint venture; otherwise, the management provisions can be included in the joint venture agreement.

Where the manager is a participant in the joint venture holding a participating interest, it is common for the manager to be in a 'no risk, no reward' position. This means that the manager will be reimbursed for costs and expenses, but is not otherwise remunerated for carrying out the role. It also means that each participant will be required, to the extent of its participating interest in the joint venture, to indemnify the manager against all losses incurred by the manager in carrying out its role. In many cases, the indemnity in favour of the manager will apply even if the manager has been negligent in carrying out its duties, although the indemnity will not usually extend to losses resulting from wilful misconduct. The rationale is that as the manager is not making a profit, it should not be required to assume potential liability for its actions.

Where the manager is a third party, participants can expect to pay for the manager's services. As such, any management agreement will usually require the manager to assume responsibility for the consequences of its actions.

Whether set out in the joint venture agreement or in a separate (third-party) management agreement, the identity of the manager and its rights, powers, duties and obligations must be clear. The circumstances in which the manager may be removed will also be an important decision for participants; as will the level of discretion to be afforded to the manager (both operationally and in terms of expenditure limitations). In particular, the joint venture agreement should carefully specify the manager's contracting limitations and expenditure authority.

As well as determining the scope of the manager's duties, joint venture participants will need to decide whether the manger is to owe fiduciary obligations to the participants and, if so, the scope of those obligations.

4.4 Management committee

Unincorporated joint venture agreements typically establish a management committee to control and direct the manager. It is usual for the management committee to be made up of one or more representatives of each of the joint venture participants. The joint venture participants must decide what level of control the management committee has and set this out in the joint venture agreement. This decision should strike the right balance between having sufficient oversight while still allowing the manager to smoothly manage the day-to-day affairs of the joint venture.

4.5 Voting rights and thresholds

The mechanism governing voting control in the management committee is a vital part of any joint venture agreement, as it is through the management committee's voting rights that the non-managing participants can exercise a level of control over the manager. First, the participants should determine those decisions essential to joint venture operations that they consider require agreement by all participants. Second, the level of agreement required to approve certain operations should be determined (eg, majority vote, unanimous consent). The joint venture agreement should set out which matters require the unanimous vote of all participants and which require a lesser threshold or majority vote. What constitutes a 'majority vote' should also be clearly set out.

A 'majority vote' need not necessarily mean a simple majority vote. It is common to see definitions along the following lines:

> *A majority vote is the affirmative decision of two or more parties, which are not related, holding more than 65% of the aggregate participants' interest of those parties then entitled to vote. Parties that are related corporations will be counted as one party, with a deemed participating interest for voting purposes of the aggregated participating interests of all of them.*

The voting provisions of the joint venture agreement should also set out a mechanism for resolving deadlocks (if a deadlock can occur having regard to the agreed voting thresholds), and the circumstances in which a participant may lose its right to vote.

4.6 Work programmes and budgets

One of the functions of the manager of the joint venture will be the preparation and submission of work programmes and budgets to the management committee for approval. It is these programmes and budgets that set the scope and boundaries of the proposed joint venture operations.

It is up to the participants to determine the level of work scope and budgetary detail that the manager should include in any proposed work programme and budget. Non-manager participants usually seek to mandate that greater levels of detail are required, because they want to limit the manager's discretion.

Some joint venture agreements require the manager to prepare an annual work programme and budget; others specify different types of work programmes and budgets that are prepared according to the activities to be carried out by the joint venture, rather than on an annual basis. Examples of the latter include:

- exploration work programmes and budgets;
- appraisal work programmes and budgets;
- development work programmes and budgets; and
- production work programmes and budgets.

It is common to see different voting thresholds for different types of work programmes and budgets – for example, a development work programme and budget involving significant capital expenditure may require a unanimous vote, whereas an exploration work programme and budget may require only a majority vote. In some circumstances, minority parties can be voted into and forced to participate in a work programme and budget even if they voted against the resolution to approve the relevant work programme and budget.

4.7 Authorities for expenditure

The ability of the manager to incur expenditure may be further controlled by the non-manager participants by including an AFE mechanism in the joint venture agreement. Such a provision requires the manager to submit a request for authority to incur expenditure identified in the approved work programme and budget before the commitment can be incurred. Joint venture agreements typically require a majority vote for approval of an AFE; however, it is common to have deemed approval for an AFE where the AFE is consistent with the underlying approved work programme and budget. The AFE mechanism is more common in petroleum joint venture agreements than in mining joint ventures, because of the relatively high expenditure commitments; but it may also be found in mining joint venture agreements.

4.8 Cost overruns

Particularly where a work programme and budget is set according to activity rather than on an annual basis, expenditure required to complete the work programme can increase from that set out in the approved budget. This can arise for reasons such as:

- a shortage or unavailability of equipment;
- a skills shortage;
- higher oil prices;
- currency fluctuations; and
- poor planning, budgeting or execution on the part of the manager.

Some joint venture agreements are restrictive: the manager's expenditure authority is strictly limited to incurring costs and expenditure as set out in the approved work programme and budget. If the manager cannot complete the work programme for the budgeted amount, it will need to obtain the approval of the management committee to incur the additional expenditure. Other joint venture agreements allow the manager to overrun the amount set out in the approved work programme and budget by up to an agreed dollar or percentage limit, and only beyond that limit will the manager need to seek further approval.

4.9 Cash calls

The cash-call mechanism enables the manager to call on participants to provide funds for project costs to be incurred by the manager. The required funds will be contributed by the participants in proportion to their respective participating interests. The cash-call provisions of a joint venture agreement need to set out when the manager is permitted to issue a cash call, the period within which the participants must pay the cash and the joint venture bank account into which the cash must be paid.

Failure to pay a cash call when due usually results in default. The joint venture agreement should also set out any interest payable on unpaid amounts and whether the non-defaulting participants have an obligation to make up any shortfall by a defaulting party. This allows joint venture activities to continue to run smoothly without putting third-party contracts into default.

4.10 Default mechanisms

The most common form of default is a failure to pay a cash call when due. Other common forms of default include:

- failing to participate appropriately (eg, failure to attend meetings);
- misusing joint venture assets in a way that is adverse to the undertaking (eg, disclosing confidential information privy to the joint venture parties); and
- encumbering assets in a manner that makes it more difficult to satisfy joint venture obligations.

The function of a default mechanism is twofold. Default mechanisms:

- act as a deterrent to default; and
- offer non-defaulting parties remedies to address the default.

There are several ways to address default in a joint venture agreement.

(a) *Loss of joint venture rights*

A joint venture agreement can be drafted so that a default can result in a loss of joint venture rights. Depending on the nature of the default, that could take the form of a loss of voting rights or suspension of the party's right to take the product of the joint venture.

(b) *Buy-out of defaulting participant's interest*

Non-defaulting parties may be given a right to buy the defaulting participant's interest in the joint venture. Care needs to be taken when drafting such provisions to ensure that:

- there is certainty around how the price payable for the interest is calculated; and
- any discount to the market value of the interest is not seen as a penalty, which may be legally unenforceable.

(c) *Enforcement of cross-charge*

Under a deed of cross-charge, each participant grants a charge in favour of each other participant and the manager to secure the payment of any sum that it is required to pay under the joint venture agreement. Each participant may charge, for example:

- its interest in its share of production;
- the proceeds of sale of its share of production; or
- its sales contracts.

The charge is usually in priority to other securities. The cross-charge allows non-defaulting participants to appoint an administrator to sell the defaulting participant's interest to recover amounts owing. The administrator will need to observe any pre-emptive rights.

(d) *Dilution of defaulting party's interest*

Dilution is a fairly common default mechanism by which a participant's interest in the joint venture is diluted or reduced and the other participants assume a greater interest in the joint venture. This mechanism allows the defaulting party to remain in the joint venture, albeit with a smaller interest, allowing operations to continue to run smoothly rather than having an outright sale process. If dilution is adopted, the joint venture agreement will need to contain a mechanism or formula for agreeing on the percentage of dilution. Dilution assumes that non-defaulting participants have an appetite to assume a greater interest in the joint venture.

(e) ***Forfeiture of defaulting party's interest without compensation***

Forfeiture is a mechanism that should be used carefully. As a default remedy, forfeiture involves a participant forfeiting its entire interest in the joint venture. From a legal perspective, forfeiture can amount to a penalty and can be unenforceable. For a commercial perspective, forfeiture may also present difficulties, such as:

- difficulty in securing funding from lenders concerned about the risk of forfeiture;
- the defaulting participant's ability to pay being dependent on maintaining its joint venture interest; and
- assuming other participants have an appetite for a greater interest due to the proportionate increase in each of the non-defaulting participant's interest.

4.11 Assignment and pre-emption

A joint venture agreement should set out any restrictions that are to apply if a participant sells all or part of its participating interest in the joint venture.

In some circumstances, the restriction may simply be that a participant needs to obtain the prior written consent of the other participants (which may be subject to a reasonableness test, or a proviso that the consenting party cannot withhold consent if the proposed incoming participant is technically and financially capable of meeting its obligations under the joint venture agreement).

In other circumstances, a participant cannot sell its participating interest in the joint venture to a third party without first offering the other participants the opportunity to acquire the interest. These rights, held by the participants, are known as pre-emptive rights, or first or last rights of refusal. The sale of the interest can be stopped if the pre-emptive rights procedure outlined in the joint venture agreement is not strictly followed.

The main disadvantage of pre-emptive rights is the potential to decrease the value of joint venture interests and opportunities for sale. For example, a third party may not want to go through the often complex and costly process of making an offer if it is likely that a participant may exercise its pre-emptive rights. However, pre-emptive rights also allow the joint venture participants to control who may become party to the joint venture. Current participants will have the first opportunity to increase their interest in a project that they have a history with and understand, instead of exploring for alternative projects.

The application of a pre-emptive right can be one of the more contentious issues that may arise during the life of a joint venture. Therefore, it is vital for the mechanics of the relevant clause to reflect, insofar as possible, the intentions of the parties from the outset. The drafting of the pre-emptive rights clauses is notoriously poor and industry standard clauses contain many traps.

At the first stages of the formation of a joint venture and the negotiation of a joint venture agreement, it is reasonable for joint venture participants to focus on the rights of pre-emption as they apply to others. In fact, exiting the joint venture can be the furthest thing from the minds of the participants when documenting the joint venture arrangements. However, whether the focus is on controlling the participants with which a participant is in a joint venture or on ensuring that a participant can easily exit the joint venture for value and without difficulty, recent case law suggests that the days of circumventing the provisions of a pre-emptive rights clause may be numbered. In the meantime, while the issue is not completely settled, carefully thought-out drafting can be used to prevent circumvention of the pre-emption clause.

(a) *Poorly drafted pre-emptive rights clause*

Table 1 sets out a commonly seen, poorly drafted, pre-emptive rights clause and highlights some of the traps with this drafting.

Table 1. Poorly drafted pre-emptive rights clause

Common clause	Likely trap
a) If a party receives a *bona-fide* offer for assignment of its Sale Interest, it must give notice to the other parties of the Selling Party's Offer, including terms and conditions of and consideration offered for the sale.	What constitutes a *bona fide* offer? What if it is not *bona fide*? What happens where the *bona fide* offer is for more assets than the joint venture interest (ie, a package sale)? Does the offer need to be capable of acceptance?
b) Within 20 days, any Party (a Pre-empting Party) may give notice to each other Party that it requires the Selling Party to assign its Sale Interest to the Pre-empting Party on the same terms and conditions as the Selling Party's Offer. On receipt of such notification,	What if the pre-empting party cannot match the offer? (If the consideration offered by the third party is the issue of shares in the third party, then the joint venture participants cannot acquire the sale

continued on next page

Common clause	Likely trap
the Selling Party must assign its Sale Interest to the Pre-empting Party (or if more than one Party exercises pre-emptive rights under this clause, to each Pre-empting Party in proportion with its respective percentage interest share).	interest on the same terms and conditions as the selling party's offer.)
c) If no Party gives notice in accordance with clause (b), the Selling Party may assign its interest to the third party in accordance with the Selling Party's Offer, provided that the assignment is completed within 180 days after the 20-day period referred to in clause (b).	
d) If the Selling Party is unable to complete the assignment within the 180-day period set out in clause (c) and it still wishes to assign the Sale Interest, it must give notice to the other Parties pursuant to this clause as if no previous Selling Party's Offer had been despatched, and clauses (b) and (c) of this clause will apply.	What if the sale is stalled and a party subsequently decides to pre-empt and is not completed within 180 days after the 20-day period referred to in (b)?

* *Clauses have been extracted from previously drafted joint venture agreements. All defined words (eg, 'Party', 'Pre-emptive Party') have been defined in those joint venture agreements and are not defined in this text.*

4.12 Rehabilitation clauses

As a number of tier one mining assets in key mining jurisdictions are reaching the end of their production life, remediation, rehabilitation and the handing back of the mining tenure are now front of mind in the industry.

In key mining jurisdictions such as Australia, Canada and the United States, legislation enables the corporate veil to be pierced in circumstances where there are significant rehabilitation costs to be incurred as a result of a company's mining activities.[7] Although such legislation provides safeguards for

7 D Proudman and D Ross, AMPLA October 2017 conference paper.

circumstances where mining has caused unexpected disturbances to land, it cannot be guaranteed that those responsible will have the financial resources to rehabilitate the disturbed land to a suitable standard.

Across these jurisdictions, there are also legislative mechanisms and financial security schemes in place to govern rehabilitation and remediation practices.

In case of the sale of a later stage brownfield project, there is increased use of a rehabilitation funding agreement between the buyers and the sellers of the mine where funds are allocated by either the seller or buyer (depending on the extent of the environmental liability) to the long-term rehabilitation of the mining tenement. We have also seen the development of better contractual provisions during the formation stages of a joint venture to clearly detail mine closure and rehabilitation.

The Australian Mining Petroleum Law Association (AMPLA) is an Australian organisation for lawyers practising in energy and resource law. AMPLA has provided a Model Mining Joint Venture Agreement which acts as a template when negotiating the joint venture agreement. A similar template is provided by the Rocky Mountain Mineral Law Foundation.

The AMPLA model template includes a provision for "Rehabilitation of Joint Venture Property" which sets out obligations on the manager to propose to the management committee a proposed programme and budget for rehabilitation. The model also provides an alternative and optional clause that may be drafted into a joint venture agreement for "Security for Mine Closure and Abandonment Costs", which requires participants of the joint venture to execute a security agreement. In Canada, the joint venture agreement will usually incorporate a sinking fund for rehabilitation after the initial exploration stage of the project.[8]

One advantage of contracting a process for rehabilitation into the joint venture agreement either in advance or in the early stages of the joint venture is that joint venture participants and the management committee can plan well ahead of the mine closure.

The rehabilitation of joint venture property provisions focuses on ensuring that there are adequate funds for the manager to pay and discharge rehabilitation obligations during mine closures by requiring the participants to provide a percentage of their share for any mine closure and abandonment costs. However, the determination of mine closure and abandonment costs is a matter that will generally require a unanimous vote.

---

8 S Potter, BE Abraham and RC Johnson, "Who's on Top? A Comparison of Key Provisions of the Australian, Canadian, and U.S. Versions of Mining Joint Venture Agreements" (2010) *Rocky Mountain Mineral Law Foundation* 56.

Aboriginal law

Arend Hoekstra
Thomas Isaac
Cassels Brock & Blackwell LLP

Aboriginal rights differ around the globe and reflect different histories, legal systems, peoples and governments. Canada has long led the world in addressing Aboriginal rights through its legal regime, including as conflicts have arisen between Aboriginal peoples and mining proponents. Courts in other commonwealth jurisdictions have informed their analysis of Aboriginal law with Canadian jurisprudence and social experience.

In 2007 the General Assembly of the United Nations adopted the United Nations Declaration on the Rights of Indigenous Peoples, addressed at the end of this chapter, creating a new benchmark for the treatment of Indigenous peoples globally.

While the United Nations Declaration on the Rights of Indigenous Peoples has outlined a shared understanding of the importance of Aboriginal rights, global experience regarding how Aboriginal rights can in practice be reconciled with national legal systems (including mining regimes) remains very limited and originates predominantly in Canada. This chapter examines Aboriginal rights from the Canadian perspective, including Canada's decades of experience in addressing Aboriginal rights in a mining context.

1. Introduction

Aboriginal rights and treaty rights are applicable throughout all of Canada. These rights have been recognised since the earliest British and French settlers arrived in Canada's east, and were given constitutional protection in the early 1980s in Section 35 of the Constitution Act 1982.[1]

In the decades since, the jurisprudence regarding Aboriginal and treaty rights has developed and evolved. With more than 60 decisions from the Supreme Court of Canada addressing Section 35 and Aboriginal and treaty rights, Canada's legal regime has developed tools and mechanisms for balancing the rights and societal interests of Aboriginal peoples with broader national interests. These mechanisms allow for mining and economic development, while protecting Aboriginal and treaty rights.

1 Constitution Act 1982, being Schedule B to the Canada Act 1982 (UK), 1982, c 11.

Companies seeking to explore for, develop or operate a mine in Canada should inform themselves about Canada's unique legal regime for the protection of Aboriginal and treaty rights. Unprepared mining proponents risk potentially indefinite delays, legal and physical obstruction, the rescission of governmental and regulatory approvals, and lost goodwill.

This chapter provides an overview of:

- the constitutional protection afforded to Aboriginal peoples;
- Canada's Aboriginal peoples;
- Aboriginal and treaty rights;
- Aboriginal interests in land;
- the Crown's duty to consult, including obligations on private parties; and
- statutes that regulate relationships between Aboriginal peoples and the resource sector.

2. The Crown in Canada

Understanding Aboriginal law requires an understanding of the basis for law in Canada. The actions and laws of governments in Canada are, in theory, the actions and laws of the queen, otherwise referred to as the 'Crown'. The 'Crown' in Canada can describe either the federal or provincial governments and is sometimes referred to as 'the Crown in right of Canada' or, in the case of the province of British Columbia, 'the Crown in right of the province of British Columbia'. The Crown in all manifestations in Canada is sovereign, restricted only by the Constitution of Canada, which divides power between federal and provincial governments, and restricts how the Crown may impact on the rights of Aboriginal peoples and Canadians generally.

3. Constitutional protection

In 1982 Aboriginal and treaty rights were recognised and affirmed in the Constitution of Canada. Section 35(1) of the Constitution Act 1982 provides that the "existing Aboriginal and treaty rights of the Aboriginal peoples of Canada are hereby recognized and affirmed". Section 35, at its core, signalled a new approach to Crown-Aboriginal relations generally, including making reconciliation a core precept of post-1982 Canadian society and a core value in the development of Canadian Aboriginal jurisprudence.

Under Section 91(24) of the Constitution Act 1867,[2] "Indians and Lands reserved for Indians" fall within the federal government's legislative power.

Aboriginal rights and treaty rights affect both the federal and provincial Crowns. Despite Section 91(24) of the Constitution Act 1867, the Supreme Court of Canada has made it clear that when a government exercises Crown

---

2 Constitution Act 1867 (UK), 30 & 31 Victoria, c 3.

power, the exercise of that power is burdened by the Crown's obligations toward the Aboriginal peoples in question.[3]

The types of obligations owed by the Crown to Aboriginal peoples in Canada depends on the unique rights of the specific Aboriginal peoples and the specific actions of the Crown. In circumstances where governments control Aboriginal property or otherwise control interests on behalf of Aboriginal peoples, the Crown is said to have a fiduciary duty.[4] Where the Crown's actions or decisions can potentially adversely impact on an Aboriginal or treaty right, the Crown has a duty to consult with, and where appropriate accommodate, the potentially impacted Aboriginal group. When dealing with Aboriginal peoples generally, the Crown is required to act in such a way that is consistent with the principle of the honour of the Crown.

Institutions which are arm's-length from a government may also have Aboriginal law obligations. For example, depending on the powers granted to a regulatory agency by a legislature, an agency may be constrained by the Crown's duty to consult.[5]

4. Canada's Aboriginal peoples

The term 'Aboriginal peoples', as used in the Constitution Act 1982 and forming the 'Aboriginal' element of 'Aboriginal rights', refers to three distinct groups:

- First Nations (referred to in the Constitution Act 1982 as 'Indians');
- Métis; and
- Inuit.

Each group is distinct, with its own culture, heritage, language, customs, practices and beliefs – though even within a group, distinct languages, cultures, customs and practices exist.

Section 35(2) of the Constitution Act 1982 defines 'Aboriginal peoples 'as including the "Indian, Inuit and Métis peoples of Canada". Recent cases have suggested that the definition of 'Indigenous peoples' – a category potentially broader than 'Aboriginal peoples' – may be afforded certain rights, despite not having the constitutional protection enshrined by Section 35 of the Constitution Act 1982.[6] Although it is unclear what rights may be afforded to a potentially broad categorisation of 'Indigenous peoples', mining proponents would be prudent to contemplate Indigenous groups that do not fall within the definition of 'Aboriginal peoples' and to seek legal advice when planning consultation with potentially impacted peoples.

---

3 *Grassy Narrows First Nation v Ontario (Natural Resources)*, 2014 SCC 48, [2014] 2 SCR 447.
4 *Guerin v The Queen*, [1984] 2 SCR 335, 13 DLR (4th) 321.
5 *Clyde River (Hamlet) v Petroleum Geo-Services Inc*, 2017 SCC 40 at para 29, [2017] SCJ No 40 (QL).
6 *Daniels v Canada (Indian Affairs and Northern Development)*, 2016 SCC 12, [2016] 1 SCR 99; *Chippewas of the Thames First Nation v Enbridge Pipelines Inc*, 2017 SCC 41, [2017] SCJ No 41 (QL); *Clyde River, supra* note 5.

5. Aboriginal and treaty rights

Nearly all lands in Canada are subject to Aboriginal or treaty rights. Aboriginal rights differ from treaty rights. While Aboriginal rights are based on historic practices, use and customs, treaty rights are those rights set out in agreements between the Crown and Aboriginal peoples.

5.1 Aboriginal rights

Constitutionally recognised Aboriginal rights fall along a spectrum with respect to their degree of connection with the land.[7] To qualify as an Aboriginal right, an activity must be an element of a practice, custom or tradition that is integral to the distinctive culture of the Aboriginal group claiming the right.[8] Aboriginal title is a subset of Aboriginal rights which grants the right of exclusive use over land. However, even if an Aboriginal group does not have Aboriginal title to the underlying lands, they can nevertheless have Aboriginal rights. While Aboriginal title requires exclusive historic occupation (or more precisely, the intention and capacity to retain exclusive control),[9] different Aboriginal groups can hold Aboriginal rights over the same piece of land.

5.2 Treaty rights

Treaty rights are those rights provided by treaties and land settlement agreements between Aboriginal peoples and the Crown. Treaties are the formal mechanism that outline Crown-Aboriginal relations and, broadly speaking, the promises made to Aboriginal people by the Crown.[10] There are three general categories of treaties:

- the pre-Confederation treaties;
- the post-Confederation treaties; and
- modern-day treaties or 'land claim settlement agreements'.

The specific treaty determines what specific rights are held.[11]

Treaties are subject to geographic limitations, either expressly or by way of implication. Aboriginal peoples with ties to a specific treaty are permitted to exercise their treaty rights in the geographic area that the treaty encompasses. Treaty-signing peoples may not necessarily be able to exercise their treaty rights throughout an entire treaty area. Instead, the area subject to treaty rights will vary between different treaty-signing peoples, depending on where those peoples conducted the protected activity at the time the treaty was signed.[12]

---

7 *Delgamuukw v British Columbia,* [1997] 3 SCR 1010, 153 DLR (4th) 193.

8 *R v Van der Peet,* [1996] 2 SCR 507, 137 DLR (4th) 289.

9 *Tsilhqot'in Nation v British Columbia,* 2014 SCC 44 at para 47, [2014] 2 SCR 257.

10 Thomas Isaac, *Aboriginal Law,* 5th ed (Canada: Thomson Reuters Canada, 2016) at pp110.

11 *Ibid.*

12 See *Prophet River First Nation v Canada (Attorney General),* 2017 FCA 15 for an illustration of where an Aboriginal group was unable to provide sufficient evidence to support its claim of having rights over certain Treaty 8 lands.

Treaty rights are not absolute and can be limited and infringed by the Crown. For example, treaty rights can be limited by a concept called 'visible, incompatible land use'. This concept means that if privately owned land is occupied or put to a visible use, Aboriginal peoples do not have a right of access. If the land is unoccupied and not being put to visible use, Aboriginal peoples will have a right of access, pursuant to the respective treaty, to exercise their rights.[13] The Alberta Court of Appeal in *Brertton*[14] held that mining is a visible, incompatible use of the land and treaty rights do not apply.[15]

Treaties may also include infringement clauses. The post-Confederation treaties include clauses that subject treaty rights to Crown infringement. While these treaties often appear to grant the Crown broad discretion to interfere with Aboriginal rights, the Supreme Court of Canada has found that the Crown owes Aboriginal groups a duty to consult and, where appropriate, accommodate their interests when infringing treaty rights.[16] In addition, in exercising these clauses, the Crown must be in conformity with the honour of the Crown and will be subject to the fiduciary duties of the Crown that arise when the Crown is dealing with Aboriginal interests.[17]

Modern treaties or land claim settlement agreements typically include formal consultation processes. For the most part, if there is an express clause that allows for infringement or if a consultation process is specified in a modern treaty or a land claim settlement agreement, the court will give full effect to the provisions of such treaty or agreement. The court will not allow one party to renege unilaterally on its constitutional undertaking by superimposing further rights and obligations that are already provided for in the treaty.[18] Despite the content of these comprehensive modern treaties, the Crown always has a duty to consult and, where appropriate, accommodate when its actions could interfere with the rights of an Aboriginal group. Where a treaty provides a mechanism for consultation, the treaty provisions will govern and where there are gaps, the court may use general consultation principles, discussed later in this chapter, to supplement the consultation obligations.[19]

5.3 Infringement of Aboriginal and treaty rights

Aboriginal and treaty rights are not absolute and instead exist alongside principles of Crown sovereignty. A fundamental aim of Canada's developing Aboriginal law is to reconcile principles of Aboriginal and treaty rights with the British legal framework that forms the basis of Canada's constitutions.

---

13 *Aboriginal Law, supra* note 10 at pp121.
14 *R v Brertton,* 1999 ABCA 285, [1999] AJ No 1152 (QL), leave to appeal to SCC refused, [1999] SCCA No 594.
15 *Ibid.*
16 *Mikisew Cree First Nation v Canada (Minister of Canadian Heritage),* 2005 SCC 69, [2005] 3 SCR 388.
17 *Grassy Narrows, supra* note 3.
18 *Beckman v Little Salmon / Carmacks First Nation,* 2010 SCC 53, [2010] 3 SCR 103.
19 *Ibid.*

The Crown may act, enact legislation and make decisions that potentially infringe Aboriginal and treaty rights where the infringement can be justified. The Supreme Court of Canada has set out a two-part test for determining whether the potential infringement is justified:

- whether the Crown was pursuing a compelling and substantial objective; and
- whether the "honour of the Crown" was upheld when potentially infringing Aboriginal or treaty rights.[20]

The court has held that valid objectives for justifying infringement can include "development of agriculture, forestry, mining, hydroelectric development, general economic development, protection of the environment or endangered species, building of infrastructure and settlement of foreign populations".[21]

Typical questions raised when considering whether the honour of the Crown was upheld include whether:

- the potentially infringed rights were given adequate priority in relation to other rights;
- infringement of Aboriginal and treaty rights was as minimal as necessary to effect the desired results;
- there was adequate consultation with the Aboriginal peoples' concerns; and
- in circumstances of expropriation, fair compensation was available.

The "honour of the Crown" is discussed more thoroughly in section 7.

Where Aboriginal rights are unproven, the court has adopted an alternative to the infringement test: consultation. Most Aboriginal rights and some elements of treaty rights are unproven. In interacting with these rights, the Supreme Court of Canada has set out obligations for dialogue and consultation through the *Haida Nation*[22] framework. The *Haida Nation* framework imposes on the Crown a duty to consult, and where appropriate accommodate, Aboriginal peoples whenever a Crown decision or action could potentially adversely impact on an asserted Aboriginal or treaty right. This framework allows for the protection and respect of Aboriginal rights without requiring Aboriginal groups to engage in the costly and adversarial process of proving their Aboriginal or treaty rights.

---

20 *R v Sparrow* [1990] 1 SCR 1075, 70 DLR (4th) 385.
21 *Delgamuukw, supra* note 7 at para 165.
22 *Haida Nation v British Columbia (Minister of Forests)*, 2004 SCC 73, [2004] 3 SCR 511.

6. Aboriginal interests in land

Most mineral rights in Canada belong to the Crown and most privately held mineral and surface rights are acquired from the Crown. However, because Crown title is burdened by the pre-existing legal rights of Aboriginal peoples, proponents must consider how Crown title is affected by Aboriginal title in the specific area in which staking or exploration is planned.

6.1 Aboriginal title

Aboriginal title is the Aboriginal right to land. It finds its roots in the historical fact that Aboriginal peoples occupied Canada prior to the arrival of Europeans.[23] Aboriginal title is a sub-species of Aboriginal rights, which are constitutionally recognised and affirmed in Section 35(1) of the Constitution Act 1982, and is the highest form of land-based Aboriginal rights.

The Supreme Court of Canada has provided a framework for understanding the meaning of Aboriginal title within the context of Section 35 and its practical application. Where Aboriginal title is infringed, the remedies available vary depending on the rights of the respective parties, the location of the land in question and the balancing of the various societal interests involved.[24] However, the Supreme Court of Canada in *Tsilhqot'in*[25] did note that Aboriginal title includes the right to decide how land will be used, the right to the economic benefit of the land and the right to proactively use and manage the land.[26] In addition, the court said that Aboriginal title confers ownership rights similar to those associated with fee simple.[27]

(a) Nature and scope

"Aboriginal title encompasses an exclusive right to the use and occupation of land, ie, to the exclusion of both non-aboriginals and members of other aboriginal nations."[28] It also "confers ownership rights similar to those associated with fee simple, including: the right to decide how the land will be used; the right of enjoyment and occupancy of the land; the right to possess the land; the right to the economic benefits of the land; and the right to pro-actively use and manage the land".[29]

At the time of the assertion of European sovereignty, the Crown acquired the underlying title to all land. However, this Crown title is burdened by the pre-existing rights and interests of Aboriginal people, including Aboriginal

---

23 *Delgamuukw, supra* note 7 at para 189.
24 *Aboriginal Law, supra* note 10 at pp65 and 66.
25 *Tsilhqot'in Nation, supra* note 9.
26 *Tsilhqot'in Nation, supra* note 9 at para 73.
27 *Ibid.*
28 *Delgamuukw, supra* note 7 at para 185.
29 *Tsilhqot'in Nation, supra* note 9 at para 73.

title.[30] When Aboriginal title is asserted the "lands are vested in the Aboriginal group, and the lands are no longer Crown lands".[31]

Aboriginal title is a unique type of interest in land as a result of its inalienability, its source and its communal nature. Lands held pursuant to Aboriginal title are inalienable, except to the Crown. The source of Aboriginal title arises from the prior occupation of Canada by Aboriginal peoples. Aboriginal title is also held communally, which further distinguishes it from normal proprietary interests. It cannot be held by individual Aboriginal persons and instead is a collective right to land held by all members of an Aboriginal nation.[32]

Once established, Aboriginal title restricts how lands may be used. Aboriginal title lands may only be used in ways that are reconcilable with the nature of the Aboriginal group's basis of claiming Aboriginal title. For instance, lands which were sacred sites may not be disturbed permanently in a way that would be inconsistent with the sacred use. This inherent limit on the uses of land held by Aboriginal title distinguishes such title from other property interests such as fee simple. If Aboriginal peoples wish to use lands held pursuant to Aboriginal title in a manner that Aboriginal title does not allow (ie, in a manner that is inconsistent with the use of such lands by future generations), they must surrender such lands to the Crown so as to convert them into non-Aboriginal lands.[33]

(b) Proving Aboriginal title

Proving Aboriginal title is a challenging and time-consuming task for Aboriginal peoples. One particular challenge is finding evidence to support a claim for Aboriginal title broadly across traditional territories, rather than just for identifiable parcels of land associated with intensive use. Aboriginal groups seeking to establish Aboriginal title judicially must satisfy a three-part test:

- The land must have been occupied by the ancestors of the Aboriginal group claiming Aboriginal title prior to the British assertion of sovereignty;
- If existing occupation of the lands in question is being offered as proof of pre-sovereignty occupation, there must be continuity between existing and pre-sovereignty occupation; and
- At the time of sovereignty, the occupation by the Aboriginal group must have been exclusive.[34]

---

30 *Aboriginal Law, supra* note 10 at pp74.
31 *Tsilhqot'in Nation, supra* note 9 at para 115.
32 *Aboriginal Law, supra* note 10 at pp75.
33 *Aboriginal Law, supra* note 10 at pp76.
34 *Delgamuukw, supra* note 7 at para 143.

(c) *Subsurface rights if Aboriginal title is proven*

Of particular interest to the resource sector are the implications that a proven Aboriginal title claim would have on subsurface rights. It is not yet clear whether Aboriginal title, where proven, includes subsurface (mineral) rights.

Currently, Aboriginal title has been proven only in the case of *Tsilhqot'in Nation.*[35] In that case the Supreme Court of Canada stated that Aboriginal title confers ownership rights similar to those associated with fee simple, including:

- the right to decide how the land will be used;
- the right of enjoyment and occupancy of the land;
- the right to possess the land;
- the right to the economic benefits of the land; and
- the right to proactively use and manage the land.[36]

Considering Aboriginal title as analogous to fee simple (Aboriginal title is similar, but not identical, to fee simple)[37] provides limited insights into whether Aboriginal title contains mineral rights. Under common law, mineral rights are not necessarily included with fee simple; however, these rights have been historically granted when not expressly withheld from a grant of fee simple.

Regardless of whether mineral rights are included in Aboriginal title, the practice of open pit mining may be fundamentally incompatible with the nature of Aboriginal title.

(d) *Infringement*

The court has expressly encouraged the Crown and third parties to seek consent from Aboriginal peoples in order to use Aboriginal title lands.[38] However, consent is not the only possible way for mineral and mining projects to be advanced on Aboriginal title lands. The Crown can justify infringement on Aboriginal title without the Aboriginal group's consent by showing that:

- the Crown's actions were backed by a compelling and substantial objective;
- the Crown discharged its procedural duty to consult and accommodate; and
- the Crown's actions are consistent with the Crown's fiduciary obligation to the impacted Aboriginal group.[39]

The Supreme Court of Canada confirmed that infringement of Aboriginal title may be justified and is a necessary part of the reconciliation of Aboriginal

35 *Tsilhqot'in Nation, supra* note 9.
36 *Tsilhqot'in Nation, supra* note 9 at para 73.
37 *Delgamuukw* as quoted in *Tsilhqot'in Nation, supra* note 9 at para 72.
38 *Tsilhqot'in Nation, supra* note 9 at para 76.
39 *Tsilhqot'in Nation, supra* note 9 at para 77.

societies with the broader Canadian community of which they form a part.[40] One compelling and substantial objective for infringing Aboriginal title can be the development of mining;[41] however, each mine requires a unique justification analysis.

Even where mining is a compelling and substantial objective, infringement cannot be justified unless the Crown action in approving the activity is consistent with the Crown's fiduciary obligation to the group. There are two components to this fiduciary duty. First, because the Crown must respect that Aboriginal title is a communal interest to both present and future generations, incursion on title without the consent of the Aboriginal group cannot be justified if this would substantially deprive future generations of the Aboriginal group of the benefit of the land.[42] 'Substantial deprivation' is likely determined by considering the scope of the impact, the proportion of the Aboriginal title lands impacted and any specific social value in the land. The incursion on title must also be proportional, in that the Crown will go no further than necessary to achieve the compelling and substantial objective.[43]

The second part of the justification test requires that where the Crown does infringe on Aboriginal rights, the Crown must discharge its duty to consult Aboriginal peoples, as may be appropriate in the circumstances. Before undertaking any action that could potentially infringe Aboriginal rights or title, the Crown must determine the proper scope of the required consultation and whether it is sufficient. The Crown's duty to consult is discussed further below in section 7.

The third component of the justification test requires an assessment of whether the infringement is consistent with:

- the special trust-like relationship between the Crown and Aboriginal peoples; and
- the honour of the Crown.

To meet these requirements, the infringement must be both justifiable and as minimal as possible. Acting honourably may include making financial arrangements: Aboriginal title has an economic component to it and fair compensation will ordinarily be required when Aboriginal title has been infringed.[44]

If Aboriginal claims are asserted but unproven, the *Haida* framework will apply to justify infringement. In order to satisfy this part of the justification analysis, the Crown must demonstrate that it made "every reasonable effort ...

---

40 *Delgamukw, supra* note 7, cited in *Tsilhqot'in Nation, supra* note 9 at para 16.
41 *Delgamukw, supra* note 7, cited in *Tsilhqot'in Nation, supra* note 9 at para 83.
42 *Tsilhqot'in Nation, supra* note 9 at para 86.
43 *Tsilhqot'in Nation, supra* note 9 at para 87.
44 *Delgamuukw, supra* note 7 at para 169.

to inform and to consult" the impacted Aboriginal group before its rights were infringed. The duty to consult does not require that an agreement be reached; nor does it give Aboriginal peoples a veto.[45]

(e) Extinguishment

Prior to 1982 Aboriginal rights, including Aboriginal title, could be extinguished by a clear federal legislative act;[46] however, the Supreme Court of Canada has yet to find an instance of such an extinguishment.

At present, Aboriginal peoples may voluntarily surrender their Aboriginal rights, including Aboriginal title, to the Crown. Such surrenders are the basis of Canada's post-Confederation and modern-day treaties. It is unclear whether the Crown can extinguish Aboriginal title unilaterally as a matter of justified infringement. Such an infringement has never been found to have occurred; however, the court's suggestion that the "settlement of foreign populations",[47] otherwise known as homesteading, is a justified infringement suggests that extinguishing Aboriginal title may be possible.

6.2 Reserve rights

The Indian Act defines 'reserves' as lands which have been set apart by the Crown for the use and benefit of an Indian band.[48]

Historically, reserve lands were typically created through some form of treaty or agreement process between the Crown and the impacted Indian band.[49] Creating reserves within provinces requires the joint action of Parliament and the provincial Crown, while reserves created in territories (subject only to the federal Crown) require only an act of Parliament.[50] Reserves are more than simply lands set aside for a First Nation; reserves have special legal attributes and can only be created by the Crown specifying its intent to designate such land as a reserve. Before a reserve is formalised, the First Nation must also indicate its acceptance of the reserve lands.[51]

The Indian Act and other federal legislation set out precise rules regarding the exploitation of resources on Indian reserves. Under the Indian Act, it is an offence to remove any minerals or timber from a reserve without the written permission of the minister. The Indian Act also provides authority to the federal Crown to regulate the exploitation of minerals, oil, gas, and timber on reserves. In order to regulate the exploitation of resources, the federal Crown has enacted a number of regulations under the Indian Act as well as additional legislation.

---

45 *Behn v Moulton Contracting Ltd,* 2013 SCC 26, [2013] 2 SCR 227 at para 29.
46 *Delgamuukw, supra* note 7 at para 181.
47 *Delgamuukw, supra* note 7, cited in *Tsilhqot'in Nation, supra* note 9 at para 83.
48 Indian Act, RSC 1985, c I-5.
49 Aboriginal Law, *supra* note 10 at pp246.
50 Aboriginal Law, *supra* note 10 at pp247.
51 *Ross River Dena Council Band v Canada,* 2002 SCC 54, [2002] 2 SCR 816.

These enactments create a system of permits and licensing which allows the minister to grant access to particular resources situated on reserve land.

Whenever the Crown grants authorisation for third parties to use reserve lands, the Crown's fiduciary obligations are triggered. The Crown's fiduciary duty requires that the Crown ensure that reserve land is used for the benefit of an Indian band; the band's interests will be at the crux of any grants of licences or permits.[52]

In most of Canada, and with the exception of Aboriginal title lands, the Crown is not required to gain the consent of Aboriginal peoples when making decisions that impact on land use. Reserves, however, form a unique legal regime. In some cases, such as long-term authorisations to occupy a reserve under Section 28(2) of the Indian Act,[53] consent of the First Nation's band council is necessary. In every case, the Crown's formal authority to make decisions impacting on reserves is constrained by the Crown's fiduciary obligations to the First Nation. As a result, the Crown regularly considers the consent of a First Nation before granting an interest in its reserve. Even where consent is granted by a First Nation, the Crown will often still analyse the transaction to ensure that the proposed activity or arrangement is in the First Nation's best interests.

In practice, when proponents are seeking to operate on reserves, the project proponent will bear the brunt of the responsibility for consulting with, and obtaining, the Aboriginal group's consents regarding access to and tenures on traditional and reserve lands.

7. The Crown's duty to consult

"Consultation in its least technical definition is talking together for mutual understanding."[54] The Crown has a duty to consult and, where appropriate, to accommodate. This duty derives from the honour of the Crown. This duty is distinct from the fiduciary duty that is owed in relation to particular cognisable Aboriginal interests.[55] The honour of the Crown is a legal principle and, when used in the Aboriginal law context, arises from the nature of the relationship between the Crown and Aboriginal peoples. The honour of the Crown requires the Crown to act honestly and diligently regarding the promises it makes to Aboriginal peoples through treaties and other agreements, and requires the Crown always to act honourably when dealing with Aboriginal peoples generally.[56]

The duty to consult is triggered when the Crown has knowledge, real or

---

52 Aboriginal Law, *supra* note 10 at pp256.
53 Indian Act, *supra* note 48.
54 *Haida Nation, supra* note 22 at para 43; T Isaac and A Knox, "The Crown's Duty to Consult Aboriginal People" (2003), 41 Alta L Rev 49, at pp61.
55 *Haida Nation, supra* note 22.
56 *Aboriginal Law, supra* note 10 at pp350.

constructive, of the potential existence of an Aboriginal or treaty right and contemplates conduct that might adversely impact on that right.[57] Although the duty lies with the Crown, the Crown can in practice delegate the procedural aspects to fulfilling this duty to third-party proponents which are interested in engaging in projects that may impact on Aboriginal rights or title. The Crown can also authorise arm's-length regulators, such as the National Energy Board, to oversee and conduct consultation.

7.1 Consultation

The Crown's duty to consult flows from Section 35 of the Constitution Act 1982. The duty to consult rests with the federal and provincial Crown, which alone hold the duty.[58] The duty "flows from the Crown's assumption of sovereignty over lands and resources formerly held by the Aboriginal group".[59]

The Crown's duty to consult arises wherever the Crown contemplates conduct that could potentially adversely impact on an asserted right.[60] The conduct required to trigger consultation can be an action by government or a decision or authorisation which permits activities, including mining, which could potentially adversely impact on an Aboriginal or treaty right (eg, the consultation obligation may arise where a mining permit, water licence or other necessary authorisation is considered). There is also no requirement for an Aboriginal group to prove that an adverse effect will occur, or that an Aboriginal right exists; the risk of a potential harm and the potential existence of a right will suffice to trigger the duty.[61]

The extent to which the Crown has a duty to consult is context specific and depends on:

- the strength of the Aboriginal group's claim to Aboriginal or treaty rights;
- the potential impact of the proposed Crown action[62] on the asserted Aboriginal or treaty right; and
- the history of dealings between the Crown and the Aboriginal group, where relevant.[63]

As a result, the duty to consult is said to fall along "a spectrum".[64]

The Crown's duty to consult may fall on the low end of the spectrum where the claim to Aboriginal or treaty rights is weak, the right itself is limited and the

57 *Haida Nation, supra* note 22.
58 *Rio Tinto Alcan Inc v Carrier Sekani Tribal Council,* 2010 SCC 43 at para 36, [2010] 2 SCR 650.
59 *Aboriginal Law, supra* note 10 at pp357.
60 *Rio Tinto, supra* note 58.
61 *Ibid.*
62 *Ibid.*
63 *Mikisew, supra* note 16 at para 63.
64 *Haida Nation, supra* note 22 at para 43.

potential infringement of the right by the Crown's actions is minor. "In such cases, the only duty on the Crown may be to give notice, disclose information, and discuss any issues raised in response to the notice."[65]

The Crown's duty to consult may fall on the high end of the spectrum where there evidence that an Aboriginal or treaty right is strong, the potential impact of the government action is likely significant to the right at issue and the Aboriginal group, and there is a high risk of non-compensable damage. At this end of the spectrum, deep consultation is required and "may entail the opportunity to make submissions for consideration, formal participation in the decision-making process, and provision of written reasons to show that Aboriginal concerns were considered and to reveal the impact they had on the decision".[66]

The duty to consult requires engagement and feedback. Through dialogue with the potentially impacted Aboriginal group, the Crown discloses the potential elements of its proposed action or authorisation.[67] During the ensuing dialogue, the Crown informs itself of the concerns of the potentially impacted Aboriginal group. With this new information, and as a consequence of the principle of the honour of the Crown, the Crown may be required to amend its actions or authorisations or otherwise accommodate the interests of the potentially impacted Aboriginal group.[68]

In order for consultation to be effective, both the Crown and the Aboriginal group must act in good faith[69] and must act to advance their respective rights in a prompt and conciliatory way.[70] Aboriginal peoples have a reciprocal duty to collaborate; the consultation process cannot be frustrated by the refusal of either party to meet or to participate.[71] Where Aboriginal groups fail to engage with the Crown in consultation, they miss an opportunity to inform the Crown of their concerns and potentially trigger the honour of the Crown, including the Crown's potential duty to accommodate their interests.[72]

In many cases, resource activities occur where there are overlapping claims for Aboriginal and treaty rights supported by credible evidence. In cases involving more than one Aboriginal group, the Supreme Court in British Columbia has recommended negotiations with umbrella organisations that are authorised to negotiate on behalf of an aggregate of Aboriginal groups.[73]

---

65 *Haida Nation, supra* note 22 at para 43.
66 *Haida Nation, supra* note 22 at para 44.
67 *Mikisew, supra* note 16 at para 64.
68 *Prophet River, supra* note 12 at para 49.
69 *Behn, supra* note 45 at para 29.
70 *Canada (Attorney General) v Long Plain First Nation*, 2015 FCA 177 at para 159, 388 DLR (4th) 209.
71 *Brokenhead Ojibway Nation v Canada (Attorney General)*, 2009 FC 484 at para 42, [2009] FCJ No 608 (QL).
72 *Sapotaweyak Cree Nation v Manitoba*, 2015 MBQB 35.
73 Sandra Gogal, Richard Riegert and Joann Jamieson, "Aboriginal Impact and Benefit Agreements: Practical Considerations" (2005) 43:1 Alta L Rev 129; *Kitkatla Band v British Columbia (Minister of Forests)*, [1999] 2 CNLR 156 at para 51 (BCSC).

7.2 Court review of duty to consult

As a primary legal principle directing the relationship between the Crown and Aboriginal peoples, the duty to consult is regularly subject to court review. In assessing whether the duty to consult has been met, the court will use a two-stage analysis. In the first stage of the analysis, the court will consider whether the Crown appropriately identified the level of consultation required. This element is examined on a correctness standard, meaning that the court will not give deference to the Crown's assessment.[74]

In the second stage of the analysis, the court will consider whether the consultation undertaken was adequate. This element is examined on a reasonableness standard: the consultation need not be perfect and mistakes will be tolerated as long as they do not strike at the heart of what is fair and reasonable in the circumstances.[75]

Even where the Crown has incorrectly assessed the level of consultation that is required in the first stage, the Crown, in action, may have still met the level of consultation required to satisfy the duty.[76] As a result, the Crown has an incentive to err on the side of caution and overestimate the degree of consultation required. Where the Crown has delegated consultation to mining proponents, it will typically seek a level of engagement that is significantly higher than the assessed level of required consultation in order to accommodate the risks that the assessment was wrong and that a third party conducted the consultation.

7.3 Delegation of consultation to regulators

The Supreme Court of Canada has stated that both federal and provincial governments can empower regulatory bodies to fulfil the Crown's duty to consult,[77] in whole or in part.[78] In order for a regulatory body to fully carry out the Crown's consultation obligations, the regulator must have the procedural powers necessary to implement consultation and the remedial powers, where necessary, to accommodate potentially impacted Aboriginal and treaty rights.[79] A final criterion is that the potentially impacted Aboriginal groups must be informed that the Crown is relying on the regulatory body to fulfil its consultation obligations.[80]

Where regulatory bodies lack the ability to fully discharge the Crown's duty to consult, the duty will remain, either in full or in part, depending on the powers and actions of the regulatory body. "Where the regulatory process being

---

74 *Ahousaht First Nation v Canada (Fisheries and Oceans)*, 2008 FCA 212 at para 34, 297 DLR (4th) 722.
75 *Katlodeeche First Nation v Canada (Attorney General)*, 2013 FC 458 at para 183.
76 *Aboriginal Law*, *supra* note 10 at pp387.
77 *Clyde River*, *supra* note 5 at para 21.
78 *Clyde River*, *supra* note 5 at para 22.
79 *Clyde River*, *supra* note 5 at para 34.
80 *Clyde River*, *supra* note 5 at para 23.

relied upon does not achieve adequate consultation or accommodation, the Crown must take further measures to meet its duty."[81]

7.4 Industry and third parties

The Supreme Court of Canada has expressly stated that any constitutional duty to consult rests solely with the Crown and not with private interests such as resource companies.[82] Although the Crown may delegate the procedural aspects of consultation to third-party project proponents, the Crown is always responsible for assessing the level of consultation required and the substantive aspects of the duty cannot be delegated.

In practice, both the federal and provincial Crown have delegated the responsibility of consultation to project proponents. Such delegation may be formal, as a result of the Crown stating or implying that project proponents must negotiate with impacted Aboriginal peoples; or informal, where the Crown does little to advance a project until all parties, including impacted Aboriginal groups, show support.

Project proponents are often best situated to effectively consult with and accommodate potentially impacted Aboriginal groups. Project proponents are incentivised to collaborate with Aboriginal groups in order to obtain permits and avoid social, legal and physical obstruction of their project. Project proponents can also identify opportunities for amending development and operating plans to reduce the impact on sensitive areas. As the primary financial benefactors of resource development, project proponents are often best situated to share profit and extend employment opportunities to impacted Aboriginal groups.

7.5 Industry practices to assist the Crown in fulfilling its duty to consult

Even where the Crown does not delegate the task of consultation to proponents, assisting the consultation effort can be an effective risk management exercise. These efforts can build consensus, reduce project opposition and give confidence to government and regulatory decision makers.

There is no formal process for consultation and the preferred approach to consultation may vary between Aboriginal groups. Effective project proponents often reach out in advance of developing or proposing a project, to build initial relationships with the impacted Aboriginal group and establish a framework for how future consultation and partnerships will develop.

Consultation efforts typically include:

- funding traditional knowledge studies to learn more about the cultural significance of an impacted area;

---

81 *Clyde River, supra* note 5 at para 22.
82 *Haida Nation, supra* note 22 at para 43.

- conducting archaeological surveys of the project area;
- meeting with group leadership to build a relationship and discuss concerns;
- meeting with, updating and generally informing Aboriginal group leadership of project developments;
- presenting project presentations at community meetings, listening to feedback and responding to questions; and
- engaging community members to act as relationship facilitators.

A report by Natural Resources Canada, which was supported by feedback collected from energy and mining companies and Indigenous organisations on building successful relationships, identified five key success factors for engaging with Indigenous groups:

- Ensure effective communication and information sharing:
    - Communicate effectively and clearly and be prepared to engage in discussions;
    - Ensure that there is transparent and open information sharing; and
    - Ensure that there is a continuous flow of information.
- Commit to the engagement process:
    - Commit to the engagement process, follow through on it and ensure that there is follow-up.
- Build capacity:
    - Ensure that capacity building is done when it is required;
    - Ensure that communities are involved in the capacity-building process;
    - Build and support capacity throughout the lifecycle of the project;
    - Potentially, improve governance structures as part of building capacity; and
    - Ensure that funding is provided for capacity building.
- Use an open and flexible framework:
    - Ensure that the engagement framework is open, transparent, flexible, responsive and developed using a consistent approach.
- Develop an understanding of communities:
    - Ensure that there is an understanding of the values and expectations of the communities, their cultural and political complexities, and their socioeconomic, environmental, health and cultural situation.[83]

It has also become standard practice for proponents and impacted Aboriginal groups to enter into memoranda of understanding and impact benefit agreements, discussed further below.

83 Ravina Bains, *Opportunities for First Nations Prosperity through Oil and Gas Development,* Centre for Aboriginal Policy Studies and Centre for Natural Resource Studies (November 2013).

(a) *Memoranda of understanding*

Preliminary agreements between proponents and Indigenous communities can take many forms. Memoranda of understandings, which can go by other names, are one common form of preliminary agreement. These non-binding agreements create a legal framework for a relationship between the project proponent and the Aboriginal group and often provide, among other things, for the project proponent:

- to conduct exploration, drilling and preliminary mining activities on the land; and
- to access the land freely and without interference from the Aboriginal group.

Memoranda of understanding typically include provisions relating to environmental protections and standards, business opportunities for Aboriginal peoples and employment and training. They typically also contemplate the future negotiation of impact and benefit agreements or intermediary agreements.[84]

(b) *Impact and benefit agreements*

Impact and benefit agreements, also called participation agreements or benefit plans, are formal contracts, usually confidential, negotiated between mining proponents and Aboriginal groups to address the socio-economic and biophysical impacts that can arise from mining development. The impact benefit agreements generally address two primary purposes:

- to accommodate Aboriginal interests by ensuring that benefits and opportunities flow to the impacted Aboriginal community; and
- to address social risk factors within the community such as socio-economic and biophysical effects of rapid resource development.[85]

The mining industry has successfully used impact and benefit agreements to facilitate corporate-Aboriginal relations. Impact and benefit agreements often address negative effects that may arise as a result of a mining project and outline the responsibilities of each party to mitigate those negative impacts. Impact and benefit agreements vary from project to project, depending on local needs.[86] The particular provisions of the impact and benefit agreement will ultimately depend on the expectations and needs of each community, as well as the

---

84 William Hipwell, Katy Mamen, Viviane Weitzner and Gail Whiteman, "Aboriginal Peoples and Mining in Canada: Consultation, Participation and Prospects for Change", North-South Institute (2002), http://caid.ca/MiningCons2002.pdf.

85 Courtney Fidler and Michael Hitch, "Impact and Benefit Agreements: A Contentious Issue for Environmental and Aboriginal Justice" (2007) 35:2 *Environments Journal* 49, www.infomine.com/library/publications/docs/Fidler2007.pdf.

86 *Hipwell, supra* note 84.

predicted impact of the mine.[87] Impact and benefit agreements also detail the benefits that a project will bring to a local community, and can include guarantees regarding levels of local employment, royalty revenues, trust funds, training programmes, scholarship programmes and sub-contracting opportunities.

An impact and benefit agreement may establish an implementation committee usually consisting of representatives of the proponent, the Aboriginal group and, potentially, independent members who are jointly appointed. The main role of an implementation committee is:

- to keep regular, open communication between the parties;
- to promote compliance with the objectives of the impact and benefit agreement;
- to serve as a forum to discuss and resolve implementation problems; and
- to appoint special committees or staff for specific tasks.

Effective impact and benefit agreements often include dispute resolution mechanisms and require that these mechanisms be exhausted before resorting to more confrontational and costly enforcement measures, such as litigation.[88]

Negotiating and implementing an impact and benefit agreement can be a demanding process. Negotiation costs include travel expenses, legal and environmental expertise, and information dissemination among community members and others. Some impact and benefit agreements require that the company reimburse Aboriginal groups for the costs incurred in their negotiation. In other instances, the Crown has provided funding for impact and benefit agreement negotiations.[89] Implementation costs include travel costs for the committee members, costs of conducting environmental monitoring and costs of involving and hiring mediators where controversies arise. Impact and benefit agreements should specify how these expenses will be paid.[90]

Although impact and benefit agreements are not legally required for mining in most Canadian jurisdictions, they can be an effective way of reducing uncertainty and potential delays in developing mining projects. Impact and benefit agreements can also serve as evidence of both the occurrence and the adequacy of consultation. Under some regimes, including some land claim agreements in Nunavut and the Northwest Territories, impact and benefit agreements with impacted Aboriginal groups are precondition to developing a mine.

---

87 Irene Sosa and Karyn Keenan, "Impact Benefit Agreements Between Aboriginal Communities and Mining Companies: Their Use in Canada", (2001) Environmental Mining Council of British Columbia, the Canadian Environmental Law Association and Cooper Accion: Accion Solidaria para el Desarrollo, www.cela.ca/sites/cela.ca/files/uploads/IBAeng.pdf.

88 *Ibid.*

89 *Ibid.*

90 *Ibid.*

An example of the use of impact and benefit agreements is the Diavik Diamond Mine in Canada's Northwest Territories. The mine is a joint venture between Dominion Diamonds Corporation and Diavik Diamond Mines Inc, and has established relationships with the surrounding Indigenous groups: the Yellowknives Dene First Nation, the Lulsel K'e Dene First Nation, the North Slave Metis Alliance, the Kitikmeot Inuit Association and the Tilcho Treaty 11 Council. The impact and benefit agreement established a human resource inventory and business registry, giving the Diavik a record of Aboriginal human capital and Aboriginal-owned businesses for the purpose of preferred hiring and purchasing. Diavik committed to providing training, employment and business opportunities and formalised these commitments with local Aboriginal groups and the Northwest Territories under the Diavik Socio-Economic Monitoring Agreement.[91]

Financial components: Most modern impact and benefit agreements include a financial component. Providing financial consideration for the infringement of Aboriginal rights is consistent with the court's statement in *Delgamuukw* that "fair compensation will ordinarily be required when Aboriginal title is infringed".[92] Factors for determining the financial component of an impact and benefit agreements include:

- the nature, size and scale of the project;
- the Aboriginal rights at stake;
- the extent of the environmental impact associated with the project;
- whether any land claims settlement provided royalties in conjunction with the project; and
- whether compensation is being provided for past environmental damage.

A larger project may also impact on more than one Aboriginal group, and therefore the payment negotiated, independently or collectively, may be larger to accommodate separate expectations. Moreover, larger projects tend to have greater impacts and consequently may contain a larger financial component.[93]

While the factors listed above may be an effective guide for negotiating a financial component of an impact and benefit agreement, the appropriate value of the financial component will be determined only by necessity. Where the impact is moderate and the receipt of Crown authorisations is likely, the proponent may be able to negotiate a reduced financial component.

91 Aboriginal Affairs and Northern Development Canada, *Aboriginal Participation in Major Resource Development* (Policy and Coordination Branch, Lands and Economic Development Sector, Aboriginal Affairs and Northern Development Canada, 2012).

92 *Delgamuukw, supra* note 7 at para 169.

93 *Aboriginal Impact and Benefit Agreements, supra* note 73 at pp143.

Alternatively, where the impact is incidental, but by reason of treaty or statute, an impact and benefit agreement is required before the Crown authorisations are granted, the Aboriginal group will be in a position to negotiate greater financial payments.

Precedent agreements provide some insight into the potential scope of amounts payable pursuant to an impact and benefit agreement and the duty to consult generally. However, each situation is different, in terms of both the rights at risk and the leverage available to the Aboriginal group in negotiations.

On example of negotiated financial payments is the Pacific Northwest liquefied natural gas (LNG) project. The Pacific Northwest LNG project was proposed by a consortium led by the Malaysian state-owned petroleum firm, Petronas. The project, eventually abandoned due to economic and regulatory challenges, included a liquid natural gas processing facility and associated port. The British Columbia government, seeking to advance the project, entered into an agreement with the Kitselas First Nation whereby it gave financial benefits worth up to C$13 million and a 1,227-hectare land grant in exchange for the Kitselas First Nation's support of the project. By agreement, the Lax Kw'aalams First Nation was to be provided with C$98 million at various stages of the LNG development. Similarly, the benefits agreement with the Metlakatla First Nations was set to provide C$46 million in cash and Crown land.[94]

7.6 Remedies for failure to fulfil the duty to consult

There is no standard remedy for Aboriginal groups where the Crown's duty to consult is not fulfilled. The Crown's failure to consult can lead to a number of different remedies depending on the situation, including injunctive relief against the proposed activity and the suspension of authorisations until further consultation is conducted.[95]

8. Applicable legislation and declarations

8.1 Indian Mining Regulations

The Indian Mining Regulations[96] apply "to surrendered mines and minerals underlying lands in a reserve", except in British Columbia.[97] Mines situated in British Columbia are governed by the British Columbia Indian Reserves Mineral Resources Act.[98] With the consent of the band whose reserve may encompass mineral rights, the Indian Mining Regulations grant authority to the chief, Oil

---

94 Gordon Hoekstra, "BC government signs LNG benefits agreement with First Nations" (2017) *Vancouver Sun*, http://vancouversun.com/business/energy/b-c-government-signs-lng-benefits-agreement-with-first-nations.

95 *Haida Nation, supra* note 22 at paras 13–14; *Rio Tinto, supra* note 58 at para 37.

96 Indian Mining Regulations, CRC, c 956.

97 Indian Mining Regulations, *supra* note 96 at s 3.

98 British Columbia Indian Reserves Mineral Resources Act, SC 1943-44, c 19.

and Mineral Division of the Development Branch of the Department of Indigenous Affairs and Northern Development, to invite tenders for mineral rights on reserves.[99]

Where tenders have been submitted in compliance with the terms and conditions set forth by the division chief, the division chief may grant a permit allowing the permit holder to test an area for the existence of minerals.[100] Every permit expires one year after the date upon which it was issued, but can be extended.[101] Permit holders are required to pay rental rates based on the acreage of the permit area, unless a rate is otherwise provided for in the tender or permit itself.[102] Assessment work is also required on the permit area to a minimum value which is based on either the acreage of the permit or the value provided for in the tender or permit.[103]

If the permit holder seeks to further develop the mineral resource, it must then apply for a lease from the supervisor of Indian minerals for the Oil and Mineral Division of the Development Branch of the Department of Indigenous Affairs and Northern Development.[104] The term of a lease granted under the regulations is 10 years.[105] Leases with terms longer than one year require consent of the council of the band under the Indian Act.[106]

A lease entitles the leaseholder to all minerals found within the lease area, unless otherwise provided for in the initial invitation to tender, permit or lease.[107] In addition to rental fees, the leaseholder must pay royalties on minerals obtained from the lease area at a rate of 5% on either:

- the gross revenue from the mineral output at the pithead, where the minerals are sold at the lease area before treatment; or
- the market value of the mineral output at the pithead, where the minerals are not sold at the lease area before treatment, unless otherwise provided for in the initial invitation to tender, permit or lease.[108]

8.2 First Nations Land Management Act

The First Nations Land Management Act[109] allows First Nations to opt out of 32 sections of the Indian Act relating to land management; instead, First Nations are given the power to develop their own laws governing land use, the

99 Indian Mining Regulations, *supra* note 96 at s 5(1).
100 Indian Mining Regulations, *supra* note 96 at s 5(2).
101 Indian Mining Regulations, *supra* note 96 at s 7(1).
102 Indian Mining Regulations, *supra* note 96 at s 10(1).
103 Indian Mining Regulations, *supra* note 96 at s 13(1).
104 Indian Mining Regulations, *supra* note 96 at s 17.
105 Indian Mining Regulations, *supra* note 96 at s 23.
106 Indian Act, *supra* note 48 at s 28(2).
107 Indian Mining Regulations, *supra* note 96 at s 20.
108 Indian Mining Regulations, *supra* note 96 at s 31.
109 First Nations Land Management Act, SC 1999, c 24.
110 Indigenous and Northern Affairs Canada, "First Nations Land Management Regime" (2016), www.aadnc-aandc.gc.ca/eng/1327090675492/1327090738973.

environment and natural resources, and take advantage of economic development opportunities with their new land management powers.[110]

Under this optional regime, each Indian band that signs a framework agreement with the federal Crown must also pass a land code (regulations pertaining to lands) approved by the members of that Indian band, setting out the process for dealing with their reserve land interests.[111] Under the First Nations Land Management Act and pursuant to the land code, a First Nation has the authority to manage reserve and other lands and may exercise the rights and privileges held by other land owners. These rights include the capacity to enter into contracts, hold real and personal property, expend and invest money, and become party to legal proceedings.[112] Once the land code is enacted, the revenue of a First Nation is no longer collected, received or held by the Crown for the use and benefit of the First Nation, but flows directly to the First Nation.[113] The land code also empowers the First Nation to enact laws with respect to First Nation land. These laws may govern the granting of interests and licences. The First Nation becomes responsible for managing the natural resources development, conservation, protection, use and possession of the land once a land code is in effect.[114]

8.3 Extractive Sector Transparency Measures Act

The purpose of the Extractive Sector Transparency Measures Act[115] (ESTMA) is to increase transparency and deter corruption in the extractive sector by requiring extractive entities active in Canada to publicly disclose, on an annual basis, specific payments made to all governments in Canada and abroad.[116] The information contained in these reports will ultimately be publicly disclosed as required under ESTMA. ESTMA explicitly states that it applies to Aboriginal governments and bodies established by Aboriginal governments; however, whether an Aboriginal group qualifies as an 'Aboriginal government' for the purposes of ESTMA will require a unique factual analysis.

8.4 United Nations Declaration on the Rights of Indigenous Peoples

In 2010 Canada endorsed, with caveats, the United Nations Declaration on the Rights of Indigenous Peoples (UNDRIP).[117] In 2016 this support endorsement became unqualified.

UNDRIP, as written, purports to grant more expansive rights to Indigenous

111 First Nations Land Management, *supra* note 109 at s 6(1).
112 *Aboriginal Law, supra* note 10 at pp223.
113 First Nations Land Management Act, *supra* note 109 at s 19.
114 First Nations Land Management Act, *supra* note 109 at s 20.
115 Extractive Sector Transparency Measures Act, SC 2014, c 39, s 376.
116 Natural Resources Canada, "Extractive Sector Transparency Measures Act (ESTMA)" (2017), www.nrcan.gc.ca/mining-materials/estma/18180.
117 United Nations Declaration on the Rights of Indigenous Peoples, GA Res 61/295, UNGAOR, 107th plenary meeting (13 September 2007).

peoples than currently provided under Canadian law. Article 19 of UNDRIP recognises free, prior and informed consent, while Article 26 states that Indigenous peoples have the right to own, use, develop and control the lands, territories and resources that they traditionally occupied.

In many ways, UNDRIP is inconsistent with current Canadian jurisprudence.[118] For example, on its face, the duty to consult under UNDRIP is broader, though lacking any legal context, than that elucidated in Canada's legal regime, which provides a spectrum of obligations for consultation, but which, where rights are unproven, is unlikely to require consent from the impacted Aboriginal peoples, as discussed above. Similarly, UNDRIP's stipulation that lands be returned to Indigenous peoples, with no other material legal context, is inconsistent with Canada's existing legal framework and case law relating to Aboriginal title.[119]

UNDRIP is a blunt tool not easily connected to Canada's existing, sophisticated and rapidly developing Section 35 legal framework for recognising and protecting Aboriginal and treaty rights. It is not clear how UNDRIP could be incorporated into Canadian law without significantly altering the existing law relating to Canada's Indigenous peoples, as it includes the requirement to obtain "free, prior and informed consent before adopting and implementing legislative or administrative measures that may impact" on Indigenous peoples, as well as a stipulation that 'Indigenous peoples have the right to the lands, territories and resources which they have traditionally owned, occupied or otherwise acquired.' These broad provisions carry the potential of significantly altering existing rules around rights recognition and consultation.[120]

---

118 Inconsistent' does not in any way suggest that Canadian jurisprudence is deficient or does not provide material protections to Aboriginal peoples' constitutionally protected rights from unilateral state actions, both objectively and in comparison with other nation states that have Indigenous populations.

119 Thomas Isaac and Arend Hoekstra, "*Identity and Federalism: Understanding the Implications of Daniels v. Canada*" (2017) 81 Sup Ct L Rev 27 [*Hoekstra*], at pp39 and 40.

120 *Ibid.*

The life of a mining project: effective diagnosis of a social licence to operate

Jordi G Ventura
Ventura & Associates LLC

1. Introduction

Twenty years ago, I had the opportunity to meet and engage in many spirited interactions with Glicerio C, a prominent leader of the Communist Party of Peru – Shining Path.[1]

My most surprising and amusing interactions with Glicerio were when I taught him and two of his lieutenants how to drive a stick shift truck over several weekly lessons.

The most rewarding interactions occurred when we spoke on the phone – Glicerio from the mayor's house in the town in the valley below his village, and me from my home back in the United States or in Lima – just exchanging stories of mundane, everyday things that occurred in our lives.

The most difficult, challenging and gut-wrenching interactions with Glicerio involved the relocation of various Shining Path community members living in and around Glicerio's village. My mining client's open pit mine was to be located on land that had been occupied by the families of Glicerio and many other Shining Path members for centuries.

At the time of my dealings with Glicerio, I was a novice at negotiating with local communities, and certainly a neophyte when it came to dealing with arms-bearing members of groups such as Shining Path. I was unaware that social sustainability, sustainable development, corporate social responsibility or any other similar concepts even existed.

What motivated me at the time was simply to perform to the best of my abilities and live by the principle: just do the right thing.

I did not have a playbook, agenda, schedule or even a simple checklist to guide me through the countless and ever-changing deals I entered into with Glicerio and Shining Path, and I certainly had no idea as to how things might ultimately turn out as a result of our discussions. Even so, and without knowing it at the time, my colleagues and I were earning a 'social licence to operate'

1 Shining Path has been and continues to be viewed in a variety of ways: a communist militant group, a revolutionary movement and a terrorist organisation.

(SLO) – one that occurred under perhaps some of the most difficult and challenging circumstances imaginable and enabled me and my colleagues to obtain permission to uproot and relocate members of Shining Path from their ancestral lands so that my client could build an open pit gold mine.

I continuously documented and evaluated my interactions and dealings with Shining Path: through my own rudimentary field notes, then by means of discussions with my client's in-country employees and home office senior management and, most importantly, through both formal and generally informal meetings, conversations and other contacts involving myself, company employees and Shining Path.

Over time, my experiences with Shining Path led me to develop, modify and improve on a common-sense, basic and extremely simple approach to earning an SLO within this local community:

- gaining the acceptance and trust of stakeholders;
- respecting and understanding the views of stakeholders; and
- including stakeholders in the mining company's decision-making processes.

My simple approach to earning an SLO is and continues to be predicated on the same basic principle that has always motivated me, and which I believe underlies all essential elements in seeking to obtain sustainability: just do the right thing. The effective diagnosis of an SLO is an extremely successful tool to assess a mining company's operations against sustainability principles and, more importantly, to gauge a mining company's progress towards obtaining an SLO.

During the past few decades, there has been ever-increasing scrutiny of mining projects and how the mining industry conducts its affairs. At first, this was due, in large measure, to the legitimate insistence by local communities "that they benefit directly and in a meaningful way, as they [would] 'bear the cost' of the project that takes place in their 'backyard'".[2]

This ever-increasing scrutiny led local communities around the world to demand a meaningful voice in determining whether and under what circumstances mining projects might be built. And alongside this increased scrutiny, a number of new concepts emerged, including social sustainability, sustainable development and corporate social responsibility (CSR),[3] all of which were geared towards assisting (and monitoring) mining companies in conducting their activities in a responsible and sustainable manner.

More recently – due in large measure to even greater scrutiny of the mining industry's contributions to sustainability – a number of authors have recognised

2 S Joyce and I Thompson, "Earning a Social Licence to Operate: Social Acceptability and Resource Development in Latin America", *Canadian Mining and Metallurgical Bulletin 93*, Number 1037, February 2000, p49, available at https://oncommonground.ca/wp-content/downloads/license.htm.

3 Of course, these concepts have developed and exist in non-mining industries as well.

the need for a shift in the focus of the traditional sustainability concepts, and have proposed a new framework that moves beyond these concepts by suggesting innovative applications of old concepts, as well as the adoption of new, timely and well-thought-out frameworks to more appropriately address sustainability issues in mining.

This chapter:

- briefly discusses the traditional sustainability concepts and the new sustainability concepts;
- outlines the particulars of the SLO diagnosis; and
- demonstrates how the SLO diagnosis can quickly and cost effectively establish a solid foundation that will hopefully convince mining companies to embrace the new sustainability concepts and integrate them as part of the business of mining.

2. Traditional sustainability concepts

Sustainability in the mining industry is a paradoxical concept and achieving sustainability is an elusive goal. Even though mining projects can create new communities, generate new jobs and business activities, and build roads, schools and health clinics in remote and previously impoverished areas, the mining industry is still distrusted by many of the individuals and local communities it deals with on a day-to-day basis. And while the traditional sustainability concepts – social sustainability, sustainable development and CSR – have for years sought to resolve the sustainability paradox, at the end of the day these concepts have far too often proven to be time consuming and costly, and have often yielded results that are of little use when a mining company needs to make immediate and timely decisions.

2.1 Social sustainability

'Social sustainability' has been defined as "occur[ing] when the formal and informal processes; systems; structures; and relationships actively support the capacity of current and future generations to create healthy and livable communities. Socially sustainable communities are equitable, diverse, connected and democratic and provide a good quality of life".[4]

Social sustainability has been criticised as being the most overly broad and least understood of the ways of seeking to obtain sustainability. Indeed, "there is universal disagreement on what quality of life goals should be. Not only do nations disagree. So do their political parties, their religions, their cultures, their classes, their activists organizations, and so on… This means that social sustainability on a practical, implementable basis is undefined".[5]

---

4 Western Australia Council of Social Services, available at http://esg.adec-innovations.com/about-us/faqs/what-is-social-sustainability/.

5 Available at www.thwink.org/sustain/glossary/SocialSustainability.htm.

2.2 Sustainable development

One of the most influential and dominant interpretations of 'sustainable development', known as the 'Brundtland definition', is "development that meets the needs of the present without compromising the ability of future generations to meet their own needs".[6]

The Brundtland definition and the sustainable development agenda as a whole have also been the source of much criticism. As noted by one author, the Brundtland definition risks "unconditional advocacy of economic growth" or "growth in marginally less damaging forms".[7] The same author contends that these interpretations are possible because sustainable development is a concept "so amorphous that it might mean anything".[8]

Likewise, still another author contends that "the vagueness of the [Brundtland] definition…allows business and 'development' interests (and their government supporters) to claim that they are in favour of sustainable development when actually they are the perpetrators of unsustainability".[9]

2.3 Corporate social responsibility

In the mining industry, 'CSR' has been defined as the voluntary actions undertaken by mining companies either to improve the living conditions – economic, social, environmental – of local communities or to reduce the negative impacts of mining projects.

CSR programmes usually invest in infrastructure (eg, potable water, electricity, schools, roads, hospitals, hospital equipment, drainage repairs), building social capital (eg, providing high-school and university education, providing information on HIV prevention, workshops on gender issues, information on family planning, improving hygiene), and building human capital (eg, training local people to be employed by the mining enterprise or to provide outsourced services; promoting and providing skills on micro business, aquaculture, crop cultivation, animal rearing and textile production).

Despite the number of laudable benefits that CSR programmes provide to local communities, CSR – as is the case with social sustainability and sustainable development – is not free from criticism. There are claims that:

- CSR programmes are just part of a company's public relations strategy – they do not really intend to benefit the local community and exist simply to bolster the company's image; and

---

6 *Report of the World Commission on Environment and Development: Our Common Future.* (Oxford: Oxford University Press, 1987).

7 R Paehlke, "Environmental Politics, Sustainability and Social Science", *Environmental Politics*, Vol 10, No 4, Winter 2001, p7.

8 *Ibid.*

9 M Jacobs, "Sustainable development: a contested concept", in A Dobson (ed), *Fairness and futurity: essays on environmental sustainability and social justice*, p24 (Oxford: Oxford University Press, 1999).

10 H Jenkins and L Obara, "Corporate Social Responsibility (CSR) in the mining industry – the risk of community dependency" (2008).

- by adopting CSR programmes, a company risks making a community dependent on the mining activity for community development.[10]

3. New sustainability concepts

For many years, mining companies had little or no interaction with local communities – everything outside a mining project's area of interest was left to the government to negotiate. Then, towards the beginning of this century, social interaction evolved to a situation where corporate handouts and donations became the norm – creating a culture where local communities were almost entirely dependent on a mining company's benevolence.

Social interaction in the mining industry is once again evolving and taking on a much greater and ever-increasing role. A number of authors in the area of sustainability have embraced this latest evolution, and recently collaborated on a new book entitled *Mining and Sustainable Development: Current Issues*[11] – a timely, considered and thorough scholarly text that sets forth a number of fresh possible sustainability solutions for mining operations.

In this book, the authors propose moving beyond the traditional sustainability concepts by suggesting innovative applications of existing concepts, as well as the adoption of new frameworks to address the new sustainability concepts. The new sustainability concepts recognise the changes in social interaction in the mining industry and are "predicated on the premise that business as usual is no longer acceptable and that sustainability matters are critical for survival of the mining industry".[12]

According to the authors, in order to ensure that mining companies are contributing positively to sustainability, "due diligence mechanisms in the form of social and environmental [and human rights] impact assessments are needed".[13] The impact assessments proposed by the authors as part of the new sustainability concepts include environmental impact assessments (EIAs), social impact assessments (SIAs) and human rights due diligence, through the use of human rights impact assessments (HRIAs).

3.1 Environmental impact assessments

Since their introduction in the United States in the 1970s, EIAs have evolved and become part of major project requirements in many countries. However, the contribution of EIAs to sustainable development in connection with mining projects has not been assessed in developing countries. The authors of *Mining and Sustainable Development: Current Issues* recognise that EIAs do not have a proven track record in the realm of mining sustainability. The authors suggest

---

11 SK Lodhia (ed), *Mining and Sustainable Development: Current Issues* (New York, NY: Routledge, 2018).

12 SK Lodhia, "Mining and sustainable development", in *Mining and Sustainable Development: Current Issues* (New York, NY: Routledge, 2018), p3.

13 *Ibid.*

that EIAs, if allowed to be utilised as a pre-decision support tool, can anticipate sustainability implications "and either design them out, or mitigate them so that sustainable mining is achieved".[14]

3.2 Social impact assessments

SIAs arose alongside EIAs in the early 1970s, primarily as a regulatory tool, and over time there has been a steady evolution in how SIAs have been performed. However, in mining countries, "SIA tends to be underdeveloped, and where it is undertaken, it is often inadequate and/or undervalued".[15] Again, the authors of *Mining and Sustainable Development: Current Issues* recognise that SIAs do not have a proven track record in the realm of mining sustainability. The authors suggest that "for SIA to play a more effective role in the development of social management systems, it needs to be integrated into the business decision-making process at a very early stage in order to provide the information critical for planning and adaptive management".[16]

3.3 Human rights due diligence

In 2011 the UN Human Rights Council unanimously endorsed the UN Guiding Principles on Business and Human Rights (UNGPs). According to the authors of *Mining and Sustainable Development: Current Issues*, "[a]s a direct consequence of the UNGPs, human rights due diligence (HRDD) became a process expected of business enterprises in any sector, anywhere in the world".[17] To help with day-to-day implementation, various practical tools exist to assist companies to "walk through a specific methodology of HRDD, often by way of a HRIA process".[18]

Once again, as in the case of EIA and SIA, the authors of *Mining and Sustainable Development: Current Issues* recognise that the concept of HRIA lacks a proven track record, given that "[the] inconsistent and spotty state of practice makes it virtually impossible for affected communities, other stakeholders and observers to monitor whether HRDD is in fact put to use, and its effectiveness".[19]

The authors' efforts in *Mining and Sustainable Development: Current Issues* to move beyond the traditional sustainability concepts (social sustainability, sustainable development and CSR) by proposing the next evolutionary phase in social interaction in the mining industry through the introduction of the new sustainability concepts (EIA, SIA and HRIA) are laudable.

---

14 A Bond and A Morrison-Saunders, "Environmental impact assessment and the quest for sustainable mining", in *Mining and Sustainable Development: Current Issues* (New York, NY: Routledge, 2018), p59.

15 S Joyce, R Sairinen and F Vanclay, "Using social impact assessment to achieve better outcomes for communities and mining companies", in *Mining and Sustainable Development: Current Issues* (New York, NY: Routledge, 2018), p70.

16 *Id* at p66.

17 M Aizawa, DC dos Santos and SL Seck, "Financing human rights due diligence in mining projects", in *Mining and Sustainable Development: Current Issues* (New York, NY: Routledge, 2018), p99.

18 *Id* at p101.

19 *Id* at p106.

The new sustainability concepts can be extremely helpful – even vital – in identifying and addressing sustainability and social issues. However, as the authors of *Mining and Sustainable Development: Current Issues* recognise, the underlying impact assessments that form the basis of the new sustainability concepts do not have proven track records; and more importantly, in order to be effective, they must be integrated into the earliest pre-decision stages of a mining project, precisely at a time when mining companies – whether junior explorers or well-established senior producers – lack either the actual or budgeted funds to embrace them.

The SLO diagnosis can overcome the early pre-decision funding hurdles presented by the new sustainability concepts by establishing a solid foundation that will hopefully convince mining companies to embrace them and integrate them as part of the business of mining.

4. Effective diagnosis of a social licence to operate

4.1 Social licence to operate

The optimal outcome of sustainability – whether through the application of the traditional sustainability concepts or the new sustainability concepts – is to earn an SLO. An SLO is not a piece of paper; rather, it is the long-term acceptance of a mining operation by stakeholder communities. These communities include not only local communities, indigenous peoples and in-country governments, but also members of international communities – environmental groups, non-governmental organisations (NGOs), financial organisations, customers, shareholders and foreign governments. Inherent in the concept of an SLO is that local communities should benefit from a mining operation "directly and in a meaningful way, as they [will] 'bear the cost' of the project that takes place in their 'backyard'".[20]

Stated otherwise, by two well-known authors in the area of sustainability, "a Social Licence to Operate exists when a mineral exploration or mining project is seen as having the approval, the broad acceptance of society to conduct its activities... Such acceptability must be achieved on many levels, but it must begin with, and be firmly grounded in, the social acceptance of the resource development by local communities".[21]

If a proposed mining project is located in close proximity to indigenous peoples, a mining company will need to seek their free, prior and informed consent (FPIC) as an unavoidable step in the development of the project. Although seeking this consent is the responsibility of the host government, not the mining company, the reality is that in most cases a successful consent process will happen only if the company takes an active role.

---

20 See *supra* note 2.
21 *Id* at p52.

Preparation for obtaining FPIC can be an extremely time-consuming process and engagement with the affected indigenous peoples must start as early as possible. FPIC entails some of the greatest challenges to be encountered by any mining company; and the irony is that preparations for FPIC need to commence exactly at a time when companies are least able to take on such a daunting undertaking. To complicate matters, a badly implemented FPIC process can derail a project completely or lead to years of tension – even conflict – throughout the life of the mine.

Although addressing the particulars associated with FPIC is beyond the scope of this chapter,[22] it should be noted that SLO and FPIC are quite distinct from each other. FPIC relates specifically to indigenous peoples who have identifiable rights within the territory where the mining company plans to operate, and there are well-defined elements to achieve FPIC.

By contrast, the concept of SLO is amorphous and has no clearly defined boundaries. A local community's interests in a mining project can have many dimensions – environment, labour, human rights, safety or health – and members of the community can take many different positions on each issue.

Despite the differences between SLO and FPIC, the basic and extremely simple approach of the SLO diagnosis applies equally to both:

- gaining the acceptance and trust of stakeholders;
- respecting and understanding the views of stakeholders; and
- including stakeholders in the mining company's decision-making processes.

This simple approach is predicated on the same basic principle that underlies all essential elements in seeking to obtain sustainability: just do the right thing.

If a mining project will impact on indigenous peoples, it is imperative that the diagnosis team (defined below) responsible for the SLO diagnosis be composed of the same individuals who engage in any interactions with the local community. Not only will this approach avoid redundant efforts, thereby resulting in significant cost savings to the mining company, it will also ensure consistency in communications – demonstrating to the indigenous peoples and members of the local community that they have equal standing and value in the eyes of the mining company.

There is much at stake in seeking to obtain an SLO. Failure to do so can – and in today's world, increasingly will – lead to project delay or suspension. A quick perusal of the news over the past few years provides a number of striking examples of what can happen when mining companies fail to obtain an SLO:

22 A more thorough discussion of indigenous peoples in South America and the various concepts that impact on them, including FPIC, can be found in J Sonoda (ed), *Indigenous Rights in South America – FPIC and Other Key Issues for Natural Resource Development* (Rocky Mt Min L Fdn, 2016).

- The Guatemalan government revoked Tahoe Resources Inc's mining licence for its flagship Escobal mine, based in large measure on a legal complaint filed in July 2017 against the Guatemalan Ministry of Energy and Mines at the Supreme Court by the Centre for Legal, Environmental and Social Action, an anti-mining NGO known in Guatemala by its Spanish acronym CALAS.[23]
- The Chilean government ordered Barrick Gold Corporation to permanently close all surface facilities at its troubled Pascua-Lama project high in the Andes after a previous suspension over concerns of environmental damage and water contamination during development.[24]
- Gabriel Resources Ltd sued the Romanian government for $4.4 billion in alleged losses when the government refused to approve the long-stalled Rosia Montana gold and silver project following community protests.[25]
- Newmont Mining Corporation deferred near-term investment in its $5 billion copper-gold Conga project in Peru in 2016 due to community opposition.[26]

In order to maximise the likelihood of obtaining an SLO, there is a need for immediate, direct and local interaction. It is vitally important that a mining company know as soon as possible, as part of the preparation ahead of a mining project – whether before sending an exploration team on the ground or before making the decision to develop a deposit – what risks and challenges it faces and, more importantly, what issues matter most to the local community that will be impacted by the mining company's operations.

A mining company's vitally important need to know what risks and challenges it faces is even more compelling when viewed against the backdrop of the instantaneous global communications in today's world. Through the advent of social media – whether via Facebook, WhatsApp, WeChat, Instagram, Twitter, Skype, Snapchat, YouTube or dozens of other sites and apps – communities in even the remotest locations on the planet are connected into information systems and support networks worldwide. To complicate matters, social media has enabled individuals and groups well beyond a mining project's physical location to identify as 'stakeholders' and become a part of a local community's issues. Social media provides massively distributed and unfiltered access to almost unlimited channels of communication. Virtually anyone can

---

23 A thorough discussion of the Escobal mine shutdown can be found in "Why Central America Stays Poor", *Wall Street Journal*, 7 October 2018.

24 "Chile orders Barrick to permanently close Pascua-Lama surface facilities", *Financial Post*, 18 January 2018.

25 "Gabriel seeks US$4.4B in damages from Romania", *National Post via Factiva*, 30 June 2017.

26 "Newmont drops Conga gold project in Peru amid political, social opposition", *SNL Metals & Mining Daily*, 20 April 2016.

disseminate information, create interest, criticise a mining company's activities and create global alliances.

Succinctly stated, any missteps by a mining company – wherever they occur – can and most likely will be immediately broadcast to an audience of literally thousands of new stakeholders. The SLO diagnosis is extremely effective at providing a mining company with vitally important information concerning the risks and challenges it faces at the outset.

4.2 Effective diagnosis of a social licence to operate

There is no 'one-size-fits-all' approach when it comes to conducting the SLO diagnosis. When embarking on a new project, mining companies need to be prepared for the fact that they are entering into a pre-existing yet dynamic environment, with established histories and cultures, and often complex political, social and economic relations among groups that can be thrown into disarray by the introduction of a mining project and the development process that accompanies it. In other words, community relations can become politicised and complicated, and can lead to or exacerbate conflicts and other unanticipated outcomes. There is no easy formula for addressing these challenges, but adhering to the basic principles of the SLO diagnosis will greatly assist in meeting these challenges.

SLO diagnoses enable mining companies to quickly, cost effectively and successfully:

- identify and assess, at the outset, what actual and potential risks and challenges exist in seeking to obtain a local community's acceptance and trust; and
- develop concrete action items to minimise and/or manage the most immediate risks and challenges.

They can do so by implementing the following actions:

- engaging in off-site preparatory work in collaboration with mining company representatives;
- conducting on-site investigative work; and
- preparing a follow-up report with concrete action items and recommendations.

These actions are geared towards establishing a solid foundation that will hopefully convince mining companies to embrace the new sustainability concepts and integrate them as part of the business of mining.

The list of particulars of the SLO diagnosis set forth below is not exhaustive and is simply meant to illustrate the various aspects of the SLO diagnosis process.

(a) ***Off-site preparatory work***

In order to ensure a quick, successful and cost-effective investigation, the SLO diagnosis initially involves preparatory work to be conducted by a diagnosis team which takes place away from the proposed mining project and prior to any significant presence of the mining company on the ground.

This preparatory work consists of, at a minimum, the following:

- meeting with relevant members of the mining company's senior management team to:
  - ascertain what, if any, endeavours have already been conducted by the mining company or additional companies (mining or otherwise) concerning the SLO and related matters, and obtain any copies of all documentation related to the same;
  - discuss any particular areas of concern that the mining company may have; and
  - agree to an overall game plan that includes expected deliverables, deadlines and costs.

  It is imperative for the mining company to grant the diagnosis team unrestricted access (pursuant to an appropriate non-disclosure, confidentiality, limited engagement or similar agreement) to review company documents, policies and procedures, interview company personnel and engage in similar activities that will maximise the diagnosis team's investigative efforts and knowledge base;
- determining host country national legislation and customs that exist and are related to:
  - land access and acquisition;
  - community consultation (especially when dealing with indigenous peoples);
  - cultural heritage; and
  - local, regional and national development;
- identifying any gaps between national legislation and international standards to ensure that all appropriate procedures will be followed in connection with the diagnosis team's investigative efforts;
- gathering relevant historical, political and socio-cultural information;
- determining whether the proposed mining project is located on or in the vicinity of lands of indigenous peoples and may impact on these lands. If this is the case, separate preparation must be made, and a possible time-consuming process must be anticipated, in order to obtain FPIC;
- identifying all stakeholders – whether formally or informally involved – that are external to the proposed mining project, including governmental regulatory agencies, global NGOs, international financial institutions and international mining industry associations;
- developing a well-thought-out procedure to provide consistent

information to landowners and land users in order to obtain permission to conduct exploration and other preliminary on-site activities;
- ascertaining the identities of key community leaders; and
- selecting and hiring one or more respected, well-connected and trusted facilitators to arrange necessary interactive and informational on-site meetings. It is usually a good idea to have the facilitator(s) also serve as a liaison between the mining company and various community factions, possibly including local law enforcement or a military presence, and otherwise act as the diagnosis team's in-country representative.

(b) On-site investigative work

Once the off-site preparatory work has been completed, the diagnosis team will relocate to the proposed site of the mining project to conduct, at a minimum, the following activities:

- meeting with the hired facilitator(s) to obtain any insights concerning the local community's feelings towards mining in general, the mining company in particular, whether there are any important issues that the diagnosis team should be aware of and/or whether in-country plans need to be altered. There is one certainty when conducting mining activities of any sort, including work associated with the SLO diagnosis: one should always anticipate changing circumstances and have contingency plans in place;
- scheduling (with enough advance notice) and holding one or more interactive community meetings where interested stakeholders can learn about the mining company's plans, exchange ideas and express their opinions. If permission is granted, the meetings should be video recorded and one or more documents should be prepared memorialising the meetings and obtaining the names and signatures of all those in attendance;
- identifying all stakeholders – whether formally or informally involved – that are immediately adjacent to or will be directly affected by the proposed mining project and assessing and/or determining:
  - their different needs, interests, values and aspirations;
  - the local history of conflict or human rights violations;
  - problems encountered by current or previous exploration projects;
  - unstable or contested land tenure arrangements;
  - intra-community divisions and inter-community conflicts and alliances; and
  - access to water and other vital resources;
- physically determining the presence of indigenous peoples or vulnerable groups;
- preparing an initial community profile (to be completed in the follow-

up SLO diagnosis discussed below), which provides enough information to initiate engagement and to begin mapping the risks or impacts that the local community and/or stakeholders may face from the development of a mining project;
- verifying the existence of any and all documents and/or agreements signed among community members, government officials, NGOs or other mining companies and, if permission is granted, obtaining copies of the same;
- verifying the existence of any oral agreements or understanding and memorialising the substance of the same;
- meticulously documenting all significant activities that take place during the on-site investigative work;
- if permission is granted, taking as many pictures and video recording as much as possible; and
- avoiding making promises or commitments so as not to raise expectations (failure to fulfil commitments will complicate and very possibly terminate a project).

(c) Follow-up

It is imperative that a follow-up SLO diagnosis report be prepared as soon as practicable after the on-site visit (and, as a time and cost saving measure, perhaps the SLO diagnosis report should be initiated during the visit).

The SLO diagnosis report should contain, at a minimum, the following:
- a description of the scope and methodology employed in conducting the on-site investigative work SLO diagnosis;
- the on-site investigative work findings and conclusions (making sure that they are duly supported);
- concrete recommendations for correcting problems/issues encountered during the on-site investigative work; and
- with the input and assistance of experts in the new sustainability concepts, proposals for the adoption of key concepts.

The diagnosis team will present an advance copy of the SLO diagnosis report to relevant members of the mining company's senior management team and will schedule one or more meetings with the senior management team to discuss the key findings and action items from the report.

At the scheduled meeting(s), in addition to discussing the particulars of the SLO diagnosis report, the diagnosis team will:
- identify and assess the most immediate actual and potential risks and challenges that exist in seeking to obtain the local community's acceptance and trust;
- decide what concrete action items must be undertaken to minimise

and/or manage the most immediate risks and challenges, and establish a budget and a timeframe for completion of the same; and

- determine which key recommendations or action items should be commenced immediately, and establish a budget and a timeframe for completion of the same.

5. Qualifications of the diagnosis team

The SLO diagnosis is most effective when conducted by a diagnosis team composed of experts in different, yet complementary disciplines.

A successful diagnosis team must possess, at a minimum, the following qualifications:

- significant experience in all issues associated with SLO, especially relating to the particulars of the EIAs, SIAs and HRIAs (the impact assessments associated with the new sustainability concepts);
- fluency in the language(s) of the host country;
- familiarity with the legal system and cultural nuances of the host country and local communities;
- familiarity with the mining company's objectives, structures and operations, past successes and failures;
- significant experience conducting SLO diagnoses (or similar);
- superior analytical ability and thoroughness; and
- the ability to write – in the host country's language(s) and in English – a clear and succinct final SLO diagnosis report

Ideally, the members of a successful diagnosis team should also be knowledgeable concerning:

- international, corporate, mining, environmental, business and related law matters, including industry-specific technical expertise; and
- all matters – legal, social, political and otherwise – that may have an impact on the company's mining operations.

The most successful diagnosis teams are headed by one or more attorneys who otherwise possess the qualifications enumerated immediately above. One of the main reasons to have the diagnosis team headed by one or more attorneys is that the mining company may be able to make effective use of attorney-client privilege, which is recognised in some form in nearly all countries:[27]

- If certain conditions are met, attorney-client privilege typically protects communications between a client and its attorney in two key ways:

27 S Altschul, R Lewis and D Zaslowsky, "Attorney-Client Privilege Around the World," ACC-GNY Ethics Marathon, 1 April 2014, available at https://m.acc.com/chapters/nyc/upload/Tab-02-ACC_Seminar_Attorney_Client_Privilege_4-1-14.pdf.

  - by prohibiting the attorney from disclosing the communications; and
  - by protecting the client from being compelled to disclose the communications in legal proceedings;[28]
- Problems discovered during the SLO diagnosis can be timely corrected/ameliorated by the mining company; and
- The attorneys have the ability to review, assess and make recommendations concerning the mining company's in-country operations, including:
  - corporate structure;
  - corporate securities and ownership;
  - the legality and enforceability of mining, building and other administrative licences and permits;
  - supplier agreements;
  - key employment contracts;
  - possible litigation risks;
  - anti-corruption;
  - anti-bribery; and
  - contractual matters.

6. Application of the SLO diagnosis

To illustrate current applications of the essential element in obtaining sustainability – the SLO diagnosis – the efforts of two companies are highlighted. Although operating in the same country (Colombia) and employing different methods, each company has adopted the same common-sense, basic and extremely simple approach in its path to earning an SLO within the local communities in which it operates:

- gaining the acceptance and trust of stakeholders;
- respecting and understanding the views of stakeholders; and
- including stakeholders in its decision-making processes.

More to the point, the simple approach adopted by both companies is – and continues to be – predicated on the same basic principle that underlies all essential elements in seeking to obtain sustainability: just do the right thing.

6.1 Red Eagle Mining Corporation

According to Alejandra Naranjo L, Red Eagle Mining de Colombia's social responsibility administrator, Red Eagle is committed to strengthening the social fabric of Santa Rosa de Osos and its surrounding communities, which make up the Santa Rosa Gold Project area of influence. To accomplish this goal, Red Eagle

---

28 *Ibid.*

has implemented social programmes that target education, culture, sports and healthcare, while at the same time strengthening local institutions and promoting sustainable businesses across diverse sectors that include all community stakeholders.[29]

Naranjo, whose full-time commitment is to ensure continuous SLO at the Santa Rosa Gold Project area of interest, proudly proclaims that Red Eagle does not seek to replace government. On the contrary, Red Eagle works towards reinforcing local institutions in order to promote social inclusion. One of its well-entrenched corporate philosophies, both in theory and in practice, is to provide a hand up, not a handout.

Based on interviews with local community members, from the moment Red Eagle arrived at Santa Rosa de Osos and even before it drilled its first hole, "it immediately engaged in social responsibility by approaching the community to understand its needs, expectations and desires for a better future".[30] Red Eagle then actively involved the community in interactive town hall forums, where interested stakeholders were able to learn about Red Eagle's plans for the Santa Rosa Gold Project and exchange ideas.

This interactive communications approach has been a defining characteristic of Red Eagle's social interaction. According to Naranjo, Red Eagle has averaged three community meetings per month for the past five years.

In another interview, a youth leader in Santa Rosa de Osos provided his own perspective on Red Eagle: "I am very grateful for the work that has been done, because since they [Red Eagle] arrived in the town, their attitude has been proactive and not authoritative."[31]

Once Red Eagle identified the stakeholders that would be impacted by its operations, it then embarked on a well-thought-out, long-term social investment – one that has had an extremely positive impact on the community. "[T]his investment transforms a human being when it contributes to improve his/her education and sporting events, planting cultural, artistic and recreational seeds. It brings you closer to technology, it trains you for life, it provides guidance in connection with interpersonal and family relations; in a nutshell, it transforms your being, making you feel useful to humanity."[32]

All of Red Eagle's social practices are generated through collaboration among multiple stakeholders. It has another well-entrenched corporate policy that encourages the creation of multi-party alliances with existing entities in its area of interest (eg, local government, academic institutions, community action boards), and these alliances have a common goal to support each other in order

29 A detailed description of Red Eagle's various and laudable accomplishments at Santa Rosa de Osos is available at www.redeaglemining.com/corporate/responsible-mining/.

30 Liliana Patiño, the teacher at La Cejita Rural Educational Centre, located in Santa Rosa de Osos.

31 Julián Díaz Arango, youth representative of the Mayor's Office, Santa Rosa de Osos.

32 *Ibid.*

to improve the quality of life of the community. "Red Eagle has been like a force that has arrived and has provided order and a logical path to all of the processes that we have been doing. It has become a supremely important ally that has enabled [the community] to generate real development plans."[33]

6.2 Royal Road Minerals Limited

When asked in an interview about operating in socially and politically challenging regions such as Colombia, Tim Coughlin, president and chief executive at Royal Road, responded: "The key to being successful in this environment is becoming a part of the community and asking the community for approval and support at the early stages of the exploration process. We work with local miners and the community to identify a specific problem and come up with a solution."

When asked to provide an example of Royal Road's solutions, Coughlin stated: "We have implemented an innovative cloud capture programme where a specialised mesh captures the mists rising over the mountains from the Pacific Ocean to condense fresh clean water. One mesh can collect 600 litres of water in five days."[34]

Perhaps the most innovative approach undertaken by Royal Road to date in order to gain an SLO within a local community occurred in December 2017, when Royal Road signed a final definitive agreement with joint social and economic organisation *Economías Sociales del Común* (ECOMUN)[35] in support of Royal Road's gold and copper exploration projects in the Nariño Province in southern Colombia, and by mutual agreement elsewhere in Colombia.

The definitive agreement "aligns Royal Road with the Colombian government's post-conflict aspirations" and demands an exemplary level of social engagement and environmental stewardship from both parties, Coughlin said:

> *This agreement is historic in that it is the first of its kind since the Peace Agreement*[36] *was reached. The agreed framework is an important step in allowing Royal Road to establish and maintain broad-based community and governmental support for its operations in Nariño Province, and reaffirms our commitment to generating economic benefits locally and to high standards of social and environmental performance.*

---

33 *Ibid.*

34 "An update on Colombia's peace process and a gold explorer gaining critical mass", Mining.com, 4 July 2016, available at www.mining.com/web/an-update-on-colombias-peace-process-and-a-gold-explorer-gaining-critical-mass/

35 ECOMUN was formed by Presidential Decree on 29 May 2017, with funding and other support provided by the Colombian government and with the aim of collective and individual economic and social reincorporation of the members of the Revolutionary Armed Forces of Colombia-People's Army (FARC-EP).

36 The Peace Agreement between the Colombian government and FARC-EP was signed on 24 November 2016 and is aimed at ending five decades of armed conflict, which killed more than 200,000 people and left millions internally displaced.

Some of the key terms of the definitive agreement include the following:

- in all aspects pertaining to environmental and social risk management, Royal Road will follow performance standards that are in line with the Equator Principles;[37]
- the parties will collaborate to formalise currently illegal mining operations in the area of influence, with a specific focus on eradicating the use of mercury and assisting operators in meeting their operational, environmental and social obligations under the law;
- ECOMUN will collaborate with and assist Royal Road in obtaining the SLO necessary to carry out exploration and future development of mineral projects in Nariño Province and elsewhere in Colombia, subject to mutual agreement between the parties;
- ECOMUN will work decisively in order to obtain the support of community and indigenous leaders in the area of influence;
- ECOMUN will assist Royal Road in working with local ethnic and non-ethnic communities to enable the company to gain the support for its operations and to implement its social and community programmes; and
- in exchange for ECOMUN's performance of its obligations, Royal Road will grant ECOMUN a net smelter royalty equal to 1% of the gold and copper produced from its mining projects in Nariño Province. The royalty is allocated entirely for the benefit of local communities.

As in the case of Red Eagle, Royal Road has appointed a qualified person with the full-time commitment to ensure continuous SLO at its Colombian projects. In recognition of the importance of this role, Jenny Arias was recently promoted from Royal Road's vice president, responsible development to executive director, responsible development.[38]

7. Conclusion

I had the privilege of participating in a panel discussion at a conference in February 2018[39] with a dear friend and former boss who I had not seen in more than 20 years – Luís Felipe Duchicela, currently the global adviser on indigenous peoples for the World Bank. Luís Felipe is Ecuadorean, of Quechua descent, from the Puruhá people of Chimborazo Province in the central Andes of

37 The Equator Principles are a set of voluntary guidelines adopted by private financial institutions to ensure that large scale development or construction projects appropriately consider the associated potential impacts on the natural environment and the affected communities.

38 A detailed description of Royal Road's various policies on key issues is available at www.royalroadminerals.com/about.

39 The panel discussion, entitled "Sustainability, Social License and Related Concepts, a Panel Discussion on What Mining Companies Need to Understand", occurred at the 2018 Annual Conference of the Society for Mining, Metallurgy & Exploration in Minneapolis, Minnesota.

Ecuador, and his complete name is Luís Felipe Atahualpa Duchicela XXVIII Santa Cruz. He is the direct descendant of Atahualpa and is the ruling Inca emperor.

During some downtime at the conference, we had the opportunity not only to catch up on family matters, but also to discuss the intricacies involved between mining activities and indigenous peoples. According to Luís Felipe, "Centuries of marginalisation, repression and exploitation have rightly caused indigenous peoples to greatly mistrust mining companies. By and large, indigenous peoples, in no uncertain terms, will never believe that sustainable development for indigenous peoples can ever result from mining activities."

After a few minutes of reflection, he added: "I know for certain that there is also an increasing number of indigenous peoples who believe that sustainable development for indigenous peoples can occur from mining activities." He paused, then continued: "But mining companies need to focus on basic human interaction – getting the trust and acceptance of indigenous peoples, respecting and understanding the views of indigenous peoples, and allowing indigenous peoples to be a part of the decision-making processes of the mining companies."

His comments left me almost speechless. I never thought I would see such willingness by a representative of indigenous peoples to speak openly and honestly about issues involving mining activities and indigenous peoples, especially in a manner that presented a ray of hope for co-existence between mining companies and indigenous peoples.

As I paused to absorb what Luís Felipe had said, I remembered all of my interactions with Glicerio of Shining Path so many years ago. Clearly, Luís Felipe – the last Inca emperor – and Glicerio, a prominent leader of Shining Path, are worlds apart when it comes to education, world views and political beliefs. And yet they are both strong leaders of their people and staunch defenders of the rights of their respective communities.

Perhaps more importantly, they are both simply human beings, just like myself, who have become my friends; friends who have taught and reminded me that the path to earning an SLO within a local community is to follow a common-sense, basic and extremely simple approach – the same approach I have mentioned throughout this chapter.

As Luís Felipe and I gave each other a departing *abrazo*, just like the many *abrazos* Glicerio and I had exchanged so many years ago – I couldn't help but think that the opportunity for mining companies to co-exist and benefit mutually with local communities, indigenous peoples and even arms-bearing groups such as Shining Path has been there all along.

Perhaps it all boils down to nothing more than living by the principle: just do the right thing.

The emerging international framework for mineral development

Luke Danielson[1]
Sustainable Development Strategies Group

1. Introduction

1.1 The context and the question

The world economy has globalised and world markets for most minerals and mineral investment are overwhelmingly dominated by private capital or by state companies that in most respects act like private companies.

In general, there are no international institutions of government that regulate the minerals markets in the same sense that national governments regulate their domestic markets.

In the absence of any kind of global regulatory system, experience shows that the purely commercial elements can tend to dominate markets, and that more diffuse interests – such as protecting environmental quality, avoiding untoward social impacts or protecting the poor or relatively powerless (eg, those who are displaced by mineral development) – can sometimes be, at best, a secondary concern.

In the more advanced economies, these externalities and the imbalance of power have eventually been addressed to some extent by domestic legislation, as those injured by the processes of development have been able to generate political movements and amass enough power to establish labour standards, environmental laws and systems for protecting the interests of the poor. They have also succeeded in ensuring that some portion of the earnings of mineral production goes to create some kind of 'safety net,' or to build better education and health, and otherwise ensure that there is broader sharing of the benefits of development. These may in the view of many still be inadequate, but it is hard to deny that government regulatory systems have led to considerable improvements since the early days of the Industrial Revolution or the mines of 50 years past.

1 Some of the material in this chapter is taken from the course that Professor Danielson teaches on this subject, available online from the Global Mining Law Center at the University of Arizona's James E Rogers College of Law. www.sdsg.org/wpcontent/uploads/2010/02/October-2016-Danielson-CV.pdf.

At the global scale, however, it has not been clear how to drive better social and environmental performance in the absence of a global government or global institutions with the power to make and enforce decisions. The major concerns are twofold: the lack of capacity of government in the poorest countries and the inability of any individual government to establish effective standards in the face of global competition.

It took the horror of World War I to press national leadership to create the League of Nations institutions.[2] These proved too weak to make and enforce effective decisions, or to prevent the outbreak of World War II, which proved an even more appalling global trauma.

The end of World War II led to the creation of a series of global institutions – the United Nations, the World Bank and the International Monetary Fund (IMF)[3] – that have had some successes. The level of armed conflict in the world has been in steady decline since the end of the Korean War – at least until very recently.[4] The length of the peace among the major European powers is unprecedented.

However, these institutions have shown serious inadequacies in dealing with the globalisation of world markets, environmental protection, human rights and the protection of the interests of the poor. Indeed, poorly managed development projects of the World Bank[5] and poorly conceived IMF policies[6] have sometimes generated their own sets of environmental and social problems. That said, these institutions have provided a framework of global standards and policy that has served the world population far better than what would have obtained in their absence.

Since the 1940s, one major new global institution – the World Trade Organization (WTO)[7] – has been created, along with some smaller special purpose institutions, and some important treaties have been negotiated. But all of this together has proven nowhere near adequate to deal with the

---

2 Christian, Tomuschat (1995), *The United Nations at Age Fifty: A Legal Perspective* (Martinus Nijhoff Publishers) p77; "Covenant of the League of Nations", The Avalon Project.

3 The World Bank and the International Monetary Fund – the so-called 'Bretton Woods institutions' – were set up in 1944 by 43 countries in Bretton Woods, New Hampshire. Their purpose is to foster global economic interaction as a way of maintaining international peace and security. The IMF coordinates monetary policies of participating states; the World Bank (as is discussed below) started by providing funds to nations seeking post-war reconstruction. These institutions have provided footing for international trade and cooperation. The United Nations, established in 1945, has chartered a more holistic approach to work towards international peace and security. It is primarily concerned with peacekeeping, conflict prevention, human rights protection, delivering humanitarian aid, promoting sustainable development and upholding international law.

4 "Fatalities in the world's conflicts declined for a second successive year in 2016, to 157,000, from 167,000 in 2015 and 180,000 in 2014." International Institute for Strategic Studies, *The ISSS Armed Conflict Survey 2017* (Routledge 2017) at 5. See also D Cortright, C Seyle and K Wall, *Governance for Peace: How Inclusive, Participatory and Accountable Institutions Promote Peace and Prosperity* (2017).

5 Environmental Issues in World Bank Projects, Independent Evaluation Group http://lnweb90.worldbank.org/oed/oeddoclib.nsf/DocUNIDViewForJavaSearch/EA662862CF792B7D852567F5005D84C7. See B Rich, *Mortgaging the Earth* (1994).

6 Larry Elliot, "The World Bank and the IMF won't admit their policies are the problem", *The Guardian*, www.theguardian.com/business/2016/oct/09/the-world-bank-and-the-imf-wont-admit-their-policies-are-the-problem.

7 World Trade Organization: "Who we are", www.wto.org/english/thewto_e/whatis_e/who_we_are_e.htm.

environmental and social externalities generated by the global economy. Among the significant problems are the following:

- lack of resources and capacity, and democratic accountability, of national governments – a problem that is most evident in developing countries, but hardly limited to them. Global trade and investment bring extraordinary levels of impact to regions where government has inadequate means to manage and control them – an issue of obvious importance in the minerals industries. Government, particularly in remote rural areas where minerals companies often operate, is sometimes simply unable to fulfil its responsibilities;
- economic competition among the largest global economies, which restrains their desire or ability to control the activities of their companies abroad;[8] and
- the 'race to the bottom',[9] in which poor performance results in cost advantages to the poor performers and economic obstacles to those trying to improve, driving performance lower and lower.

In the face of these realities, there have been some notable efforts to create effective systems of rules, to try to create more efficient, transparent markets that internalise an increasing portion of the most problematic environmental and social costs. This is the challenge of trying to create 'governance without government'.[10] It is one of the great global challenges of our era – something that is of vital importance to the mining industry, the principal subject of this chapter.

---

8 There are examples of countries trying to set some boundaries on the conduct of domiciliary companies abroad, such as the Canadian Office of the Extractive Sector Corporate Social Responsibility Counsellor, www.international.gc.ca/csr_counsellor-conseiller_rse/index.aspx?lang=eng, or the conflict minerals provisions of the US Dodd Frank legislation: "Section 1502 of the US law known as the 'Dodd-Frank Act' includes a requirement that companies using gold, tin, tungsten and tantalum make efforts to determine if those materials came from the Democratic Republic of Congo (DRC) or an adjoining country and, if so, to carry out a 'due diligence' review of their supply chain to determine whether their mineral purchases are funding armed groups in eastern DRC." Business & Human Rights Resource Centre, www.business-humanrights.org/en/conflict-peace/conflict-minerals/implementation-of-us-dodd-frank-act-rule-on-conflict-minerals-commentaries-guidance-company-actions. See Periodical and other reports, 15 USCA § 78 There is also the US Foreign Corrupt Practices Act, 15 USCA § 78dd-1, which makes it "unlawful for certain classes of persons and entities to make payments to foreign government officials to assist in obtaining or retaining business". US Justice Department, Foreign Corrupt Practices Act, www.justice.gov/criminal-fraud/foreign-corrupt-practices-act; see also the EU Conflict Minerals Regulation, which, as of 1 January 2021, will require EU importers of tin, tungsten, tantalum and gold to undertake due diligence in their supply chains to ensure what they buy is sourced responsibly and does not contribute to conflict or other related illegal activities, http://ec.europa.eu/trade/policy/in-focus/conflict-minerals-regulation/regulation-explained/. China has been working on a code of conduct for Chinese companies abroad.

9 The 'race to the bottom' is a phenomenon in which national governments, especially in poorer countries, are incentivised to keep the cost of doing business as low as possible to ensure that companies do not move production elsewhere. This dissuades national governments from passing legislation or taking regulatory measures that protect environmental and social interests, and economic pressure drives standards lower. Regina E Rauxloh, "A Call For the End of Impunity for Multinational Corporations", 14 Tex Wesleyan L Rev 297. 2008.

10 See W Reinicke, *Global Public Policy: Governing without Government?* (Brookings Institution 1998); *Governance without Government, Order and Change in World Politics* (Rosenau, ed) (Cambridge 2009); *Governance without Government? The Case of the Forest Stewardship Council* (Stephen Bell, Andrew Hindmoor) (Wiley Online Library, 2011) https://doi.org/10.1111/j.1467-9299.2011.01954.x.

These issues are of importance to the mining and minerals industries in part because they do so much business across the economic divide: most of the customers for their products and investors are based in rich countries with strong and complex systems of governance, while much of their production of raw materials is based in poor countries, where governance is so frequently lacking. There are often only the thinnest threads trying to reconcile these two very different legal worlds. That is why the mining industry is often seen at the forefront of developing international standards.

Few companies are trying to establish high-tech industries or banking centres in the rural regions of the poorest countries. Yet mining companies, oil and gas operators and other resource investors often venture into these areas.[11]

This creates considerable risks, where government is not capable of ensuring environmental protection, protecting the rights of people displaced by development or controlling corruption. It creates risk for business, but also for governments, impacted communities and the environment.

It also creates some real opportunities. Mining is one of the few options that the very poor countries have available for creating the revenue streams needed to improve conditions of life for their populations. An international framework capable of channelling economic competition in directions that benefit developing countries and their people can ensure that the mining industry can be a force for human welfare. It also offers the prospect of more stable conditions of investment, something of great interest to the industry.

1.2 Background: the limits of international law

The current system of international law has its roots in the several treaties that ended the great European religious wars of the 16th and 17th centuries, generally called the Peace of Westphalia.[12] These treaties established a concept of sovereignty[13] which, with various glosses and amendments, is still the

---

11 The United Nations Human Development Index sorts countries into categories from Extremely High Human Development to High Human Development, to Medium and Low Human Development. United Nations Development Programme, *Human Development Report 2016: Human Development for Everyone* (2016). http://hdr.undp.org/en/composite/HDI. Among those listed as having Low Human Development are countries such as Nigeria, Democratic Republic of Congo, Mozambique, Papua New Guinea, Uganda, Tanzania and Guinea, which are of considerable interest to the extractive industries.

12 Gordon A Christenson, *"Liberty of the Exercise of Religion" in the Peace of Westphalia*, 21 Transnat'l L & Contemp Probs 721, 737 (2013) ("The Congress of Westphalia convened after lengthy preparations to bring the savage Christian wars to a close... on December 4, 1644, with 135 delegates... [including] theologians, philosophers, and diplomats"). "Peace of Westphalia", *Encyclopædia Britannica Online* (Encyclopædia Britannica, Inc, 2015).

13 Gordon A Christenson, *"Liberty of the Exercise of Religion" in the Peace of Westphalia*, 21 Transnat'l L & Contemp Probs 721, 722-723 (2013) ("The Peace of Westphalia applied basic terms of an earlier religious agreement, the Peace of Augsburg of 1555, to all the major European powers... [with the] foundation of a new concept in international law based on state sovereignty"). See also Amos S Hershey, "History of International Law Since the Peace of Westphalia", *The American Journal of International Law*, vol 6, no 1, 1912, pp30–69. *JSTOR*, www.jstor.org/stable/2187396. "Peace of Westphalia", *Encyclopædia Britannica Online* (Encyclopædia Britannica, Inc, 2015).

bedrock of the international system. The sovereign state in this concept is free to regulate its internal affairs,[14] without interference from other states.

But this system does not generally create clear answers to a series of increasingly important questions, such as the following:

- Where a sovereign state simply lacks the resources and capacity to ensure the most basic level of environmental protection or worker safety, or to prevent abuse of affected communities, where can it turn for help?
- Where a country is so overwhelmed by lack of capacity and corruption that it is not receiving anything like a fair return for its resources, how is it to progress?
- If government and external investors both lack concern for the rights and wellbeing of minority ethnic groups in resource-rich regions, who will safeguard their human rights?
- Where failure to internalise environmental and social costs of production gives a country or a company a cost advantage, what are more responsible countries and companies to do aside from lowering their own standards?

The mining and minerals sector has played a central role in the development of this system of international law over time. For example, one of the most important current limitations to absolute territorial sovereignty is the principle that countries have a duty to control pollution that crosses national boundaries and are liable for environmental damages caused outside their territories by activities within their territories.

Few events were as important in establishing this principle as the Trail Smelter emissions. The Trail Smelter was within Canadian territory, but it created air pollution impacts in US territory. Decades of arbitration and lawsuits over the Trail Smelter established Canada's duty to control the emissions[15] and its liability for damages.[16] The establishment of this principle may have contributed to putting an end to the practice of locating smelters near national borders[17] in order to pass environmental costs to neighbours.

In general, however, states are not much limited in the environmental

14 Gordon A Christenson, *"Liberty of the Exercise of Religion" in the Peace of Westphalia*, 21 Transnat'l L & Contemp Probs 721, 761 (2013); quoting Leo Gross, "The Peace of Westphalia", 1648-1948, 42 Am J Int'l L 28-29 (1948). ("[The Peace of Westfalia marked] man's abandonment of the idea of a hierarchical structure of society and his option for a new system characterized by the coexistence of a multiplicity of states, each sovereign within its territory, equal to one another, and free from any external earthly authority").

15 John D Wirth, "The Trail Smelter Dispute: Canadians and Americans Confront Transboundary Pollution, 1927-41", *Environmental History*, vol 1, no 2, 1996, pp34–51. *JSTOR*, www.jstor.org/stable/3985111.

16 John D Wirth, "The Trail Smelter Dispute: Canadians and Americans Confront Transboundary Pollution, 1927-41", *Environmental History*, vol 1, no 2, 1996, pp34–51. *JSTOR*, www.jstor.org/stable/3985111.

17 See, for example, Texas Commission on Environmental Quality, American Smelting and Refining Company (ASARCO) Site, El Paso, Texas, www.tceq.texas.gov/remediation/sites/asarco; Molymex II, Commission on Environmental Cooperation, SEM-00-005 (2004), www.cec.org/sem-submissions/molymex-ii

decisions they make within their borders, absent a treaty obligation. It is entirely consistent with the Westphalian system of sovereignty to expect countries to honour obligations that they have voluntarily undertaken in treaties,[18] including the agreements that create international organisations.

1.3 The need for international standards

This leads to the fundamental rationales for having a set of international standards that govern the mining and minerals industries. The suggested rationales include the following:

- Where minerals-based development strategies are successful and generate widely shared development benefits, risk to investors is reduced. International standards are seen as avoiding obvious sources of conflict and promoting stable long-term relationships between investors and host countries.
- International standards can supplement and reinforce national efforts to ensure better management of the industry, where national governments are struggling with lack of skills, lack of resources, corruption and related problems. These standards can create an environment in which it is easier for countries to protect human rights, for companies to respect human rights and for those whose rights are not respected to have some kind of remedy.[19]
- Certain practices should not form part of the global system of economic competition. Competition should not extend to seeking cost advantages based upon child labour, uncompensated displacement of communities, mismanagement of mercury or widespread bribery of public officials. Realistically, these practices will not end overnight. But we would all benefit from being on the road to their minimisation, and companies that choose to act against them should not be forced to compete against those which see them as an economic advantage.
- Finally, common standards can reduce transaction costs and facilitate trade and investment. Where everyone uses different weights and measures, markets have high transaction costs. So there is a certain efficiency if the major players in the mining and minerals marketplace use the same reserve reporting requirements, accounting standards or due diligence processes on environmental or social issues.

There is some evidence of change: that the leaders in the international mining industry have transformed the way the best companies are doing

18 Oona A Hathaway, "International Delegation and State Sovereignty. Law and Contemporary Problems", January 2008. Available at SSRN: https://ssrn.com/abstract=1009600.

19 See United Nations, *Guiding Principles on Business and Human Rights* (2011), www.ohchr.org/Documents/Publications/GuidingPrinciplesBusinessHR_EN.pdf.

business, that government has more capacity and is doing a better job of regulation, and that civil society organisations are becoming better in their watchdog role.[20] This has tremendous potential to increase social welfare in minerals dependent developing countries. And if this indeed is happening, the legal and 'soft law' arrangements discussed below may well be an important causative factor.

2. International institutions

Several sets of guidelines or rules, originating in international institutions, apply broadly to international investors or corporations doing business abroad. While these are of considerable importance to the mining and minerals industries, they are systems that apply broadly to many industries, not just to mining. Of these, the system with the greatest impact has clearly been the system of Performance Standards developed by the International Finance Corporation, a part of the World Bank Group, and the related Equator Principles, discussed below.

2.1 The World Trade Organization and its dispute settlement process

The rules of the World Trade Organization (WTO) do not apply directly to private actors, but may have considerable consequences for them. The WTO is an international body, created by governments, that regulates international trade. It commenced on 1 January 1995 under the Marrakesh Agreement,[21] which was signed by more than 120 governments in 1994 and which replaced the earlier General Agreement on Tariffs and Trade.

Where there is an allegation by one nation that the conduct of another nation is violating WTO trade rules, the general practice is that the nations attempt to negotiate a resolution. Where negotiation fails, there is a litigation process open to the parties. It results in "panel and Appellate Body reports, which are binding upon the parties once adopted by" the Dispute Settlement Body (DSB).[22]

An example of the potential effects of this system on the mining and minerals sector are the rare earths complaints brought against China by Japan, the United States and the European Union[23] – some of which failed, but the last of which succeeded.

China is the largest producer of rare earth elements in the world. By some

20 See D Franks, *Mountain Movers: Mining, Sustainability and the Agents of Change* (Routledge 2015); J Owen and D Kemp, *Extractive Relations: Countervailing Power and the Global Mining Industry* (Routledge 2017); D Humphries, *The Remaking of the Mining Industry* (Palgrave MacMillan 2015).

21 See WTO Marrakesh Agreement Establishing the World Trade Organization, WTO, www.wto.org/english/docs_e/legal_e/04-wto_e.htm.

22 See WTO, www.wto.org/english/tratop_e/dispu_e/dispu_body_e.htm.

23 See generally, Jost Wubbecke, "Rare earth elements in China: Policies and narratives of reinventing an industry", *Resources Policy* (2013). See also WTO, www.wto.org/english/tratop_e/dispu_e/cases_e/ds431_e.htm.

measure, it has at times produced over 95% of world supply.[24] These minerals are essential in the manufacture of a wide variety of new technologies, from mobile phones and solar collectors to flat screens, touch screens and a long list of advanced military technologies.[25]

In 2010 China began to restrict its exports of rare earth minerals.[26] A variety of motives were ascribed to this action, including a political dispute with Japan.[27] China's explanation was more complex and included the suggestion that it needed to control production levels in order to get better control over the widely recognised environmental problems in its rare earths industry. The environmental issues are certainly not imaginary and there is support for the idea that China's dominance of this market is as much a result of other nations' reluctance to permit the problematic processing of these materials (without incurring much higher environmental control costs) as it is any Chinese strategy.[28]

The DSB determined the case against China.[29] The general holding was that because China's export quotas could not be justified as measures taken to conserve exhaustible natural resources – a general exception to trade restrictions under Article XX of the Marrakesh Agreement – China's actions were inconsistent with and thus in violation of its international treaty obligations.

The future effectiveness of this dispute settlement process is being questioned, at least in the near term:

> *The seven-member Appellate Body of the World Trade Organization (WTO), which makes rulings on arguments between countries over trade, has already seen two judges leave, with a third going in December.*
>
> *But the US ... has blocked replacements, leaving just four judges to deal with a growing backlog of trade disputes. A minimum of three judges are needed to sign off on every appeals decision.*[30]

Another area in which WTO rules may be of importance to the mining industry is the WTO prohibition on nations creating trade barriers aimed not at products themselves, but at production processes and methods.[31] "If trade-related environmental or health measures are to be consistent with WTO rules,

---

24 Pui-Kwan Tse, 2011 *China's rare-earth industry: U.S. Geological Survey Open-File Report 2011–1042*, 11 https://pubs.usgs.gov/of/2011/1042.

25 US Geological Survey factsheet, "The Rare Earth Elements – Vital to Modern Technologies and Lifestyles", https://pubs.usgs.gov/fs/2014/3078/pdf/fs2014-3078.pdf.

26 Keith Bradsher, "China to Tighten Limits on Rare Earth Exports", *New York Times*, 28 December 2010, www.nytimes.com/2010/12/29/business/global/29rare.html.

27 Keith Bradsher, "Amid Tension, China Blocks Vital Exports to Japan", *New York Times*, 22 September 2010.

28 See generally J Klinger, *Rare Earth Frontiers: From Terrestrial Subsoils to Lunar Landscapes* (Cornell 2017).

29 Appellate Body Reports, *China – Measures Related to the Exportation of Rare Earths, Tungsten, and Molybdenum*, WT/DS431/AB/R / WT/DS432/AB/R / WT/DS433/AB/R, adopted 29 August 2014, DSR 2014:III, p. 805

30 "Trump could cause world trade system to freeze up after vetoing appointment of judges, diplomats fear", *The Independent*, 28 November 2017, www.independent.co.uk/news/world/americas/donald-trump-world-trade-dispute-system-veto-judges-appointments-global-freeze-us-diplomats-warning-a8079876.html. The United States is also threatening to leave the WTO entirely. "Trump: US will quit World Trade Organization unless it 'shapes up'", *The Guardian*, 30 August 2018.

they cannot result in discrimination between 'like' products. Therefore, the principle of non-discrimination raises two key questions: Are products at issue 'like' products? If so, is the foreign product treated less favourably ...?"[32]

A well-known case is the *Shrimp – turtle* litigation.[33] Without the use of protective technologies, shrimp fishing can result in high mortality of endangered sea turtles. The United States banned these production techniques for its own fishing industry, but then sought to extend the ban to imported shrimp. The imported shrimp were certainly a 'like' product; the trade measure was aimed not at the shrimp, but at the production methods.

The question is then whether a country's ban (or other restriction) on the importation of a mineral product, based on disapproval of the methods of its production, would be a permissible exception to the general principle under one of the provisions of Article XX. For example, can gemstones be excluded because they are produced by child labour or coltan because its production is financing warlords who are upsetting civil order? In general, the existence of an international consensus, such as that expressed in the child labour treaties, can justify the exclusion.

Of course, the WTO rules generally do not act to sanction private choices made by consumers, such as the decision not to buy coffee without a 'fair trade' label or to buy only certified wood products. For this reason, many of the certification schemes discussed later in this chapter have avoided government involvement, and are managed by private industry and civil society participants, which can make their own decisions about what to buy without triggering WTO consequences.

2.2 The Organisation for Economic Cooperation and Development Guidelines

One set of important international rules is promulgated by an association of governments, the Organisation for Economic Cooperation and Development (OECD).[34] In 1976 the OECD adopted a Declaration on International

---

31 See International Institute for Sustainable Development, *The Legality of PPMs under the GATT: Challenges and Opportunities for Sustainable Trade Policy*, www.iisd.org/library/legality-ppms-under-gatt-challenges-and-opportunities-sustainable-trade-policy.

32 "WTO rules and environmental policies: key GATT disciplines", World Trade Organization, www.wto.org/english/tratop_e/envir_e/envt_rules_gatt_e.htm.

33 Appellate Body Report, *United States – Import Prohibition of Certain Shrimp and Shrimp Products – Recourse to Article 21.5 of the DSU by Malaysia*, WT/DS58/AB/RW, adopted 21 November 2001, DSR 2001:XIII, p6481 (www.wto.org/english/tratop_e/dispu_e/cases_e/ds58_e.htm) There is also the *Dolphin – tuna* litigation. Panel Reports, *United States – Measures Concerning the Importation, Marketing and Sale of Tuna and Tuna Products – Recourse to Article 21.5 of the DSU by the United States*, WT/DS381/RW/USA and Add.1 / *United States – Measures Concerning the Importation, Marketing and Sale of Tuna and Tuna Products – Second Recourse to Article 21.5 of the DSU by Mexico*, WT/DS381/RW2 and Add.1, circulated to WTO members on 26 October 2017 [appealed by Mexico on 1 December 2017].

34 The US Mission to the OECD describes it as "a unique forum where the governments of 34 democracies with market economies work with each other, as well as with more than 70 non-member economies to promote economic growth, prosperity, and sustainable development". US Mission to the Organization for Economic Cooperation & Development, "What is the OECD?", https://usoecd.usmission.gov/our-relationship/about-the-oecd/what-is-the-oecd/.

Investment and Multinational Enterprises.[35] Under that declaration, the OECD has developed the OECD Guidelines for Multinational Enterprises.[36]

There is considerable legal literature regarding these guidelines.[37] Much of it focuses on the perceived inadequacy of the remedy afforded. Complaints under the guidelines are essentially directed to the national contact point of the home country of the company against which the complaint is made.[38] The national contact point is generally an agency or official of the home country government. It appears that some governments take the system seriously and have a more robust process; in other countries, potential complainants may find complaints to be of questionable value.[39] In the complaints process, if the issues are not resolved earlier, the case or 'instance' goes through an assessment by the national contact point and results in a final report. The process does not result in any financial or other sanction against the company; the effect, if there is one, is largely in the court of public opinion.

This complaints mechanism has been used on a considerable number of occasions. A list of cases is published by the OECD.[40] A high percentage of these instances involve mining and minerals companies. A considerable number of the complaints relate to labour practices.

One example of a complaint directed at a mining company is *Future in Our Hands (FIOH) v Intex Resources ASA*.[41] The project is in the Philippines. Because the company developing the project is Norwegian, the case was processed by the Norwegian national contact point.[42] The decision was quite critical of the company and suggested that improved stakeholder consultations and environmental assessments and information were needed.

---

35 The declaration "was first adopted by the Governments of OECD Member countries on 21 June 1976 and has been reviewed on several occasions since then (1979, 1984, 1991, 2000 and 2011)".

36 2011 Update of the OECD Guidelines for Multinational Enterprises, OECD: Better Policies for Better Lives, www.oecd.org/corporate/mne/oecdguidelinesformultinationalenterprises.htm.

37 Evaristus Oshionebo, "The OECD Guidelines for Multinational Enterprises as Mechanisms for Sustainable Development of Natural Resources: Real Solutions or Window Dressing?" 17 Lewis & Clark L Rev 545 (2013); Christian Schliemann, "Procedural Rules for the Implementation of the OECD Guidelines for Multinational Enterprises – A Public International Law Perspective", 13 German LJ 51 (2012); Ashely L Santner, "A Soft Law Mechanism for Corporate Responsibility: How the Updated OECD Guidelines for Multinational Enterprises Promotes Business for the Future", 43 Geo Wash Int'l L Rev 375 (2011); Sara Fick Vendzules, "The Struggle for Legitimacy in Environmental Standards Systems: The OECD Guidelines for Multinational Enterprises", 21 Colo J Int'l Envtl L & Pol'y 451 (2010); Christopher N Franciose, "A Critical Assessment of the United States' Implementation of the OECD Guidelines for Multinational Enterprises", 30 BC Int'l & Comp L Rev 223 (2007).

38 OECD: Responsible Business Conduct, http://mneguidelines.oecd.org/ncps/.

39 "NCPs have deviated wildly in how they have handled NGO complaints and many have been mishandled. For example, less than 10 of the 45 NGO complaints submitted as of September 2005 concluded in a manner that reasonably satisfied the complainants." Freeman *et al*, "Guide to the OECD Guidelines for Multinational Enterprises' Complaint Procedure", *OECD Watch* (2006) https://germanwatch.org/tw/oecd-gui06.pdf.

40 A database to search all specific instances (https://mneguidelines.oecd.org/database/) and an overview of cases handled from 2000-2017 (https://mneguidelines.oecd.org/Flyer-OECD-National-Contact-Points.pdf) are available.

41 See Business & Human Rights Resource Centre, www.business-humanrights.org/en/documents/concerns-regarding-intex-resources-mindoro-nickel-project-in-the-philippines-and-company-response.

42 The final statement, or decision, in this proceeding is available at www.oecd.org/daf/inv/mne/Norwegian%20NCP%20intex_final.pdf.

Another interesting case, *Cerrejon Coal*, involved allegations against BHP that it took actions to force the relocation of villagers in Colombia.[43] In that case, the Australian national contact point worked together with those of the United Kingdom and Switzerland to settle the matters raised by the complaint. All stakeholders involved were brought into a mediation process, the result of which was payment by Cerrejon of $1.8 million in indemnities and $1.3 million for the development of sustainable projects.[44]

There have also been claims involving the Marlin Mine in Guatemala,[45] against Ivanhoe Mines and Rio Tinto over the Oyu Tolgoi mine in Mongolia,[46] and against other industry players for their labour practices. In the named cases, the respective national contact point facilitated dialogue between the parties, then considered the matter to be concluded.[47]

(a) OECD Transfer Pricing Guidelines and the BEPS Dialogue

A related set of guidance is the OECD Transfer Pricing Guidelines for Multinational Enterprises and Tax Administrations, which provide guidance on the application of the 'arm's-length principle' – the international consensus on transfer pricing (ie, on the valuation, for tax purposes, of cross-border transactions between associated enterprises).[48] This has given birth to a large and complex discussion on tax equity, with a great deal of focus on the difficulties that developing countries face in collecting taxes from multinational corporations. This is sometimes referred to as the discussion of base erosion and profit shifting (BEPS).[49]

It seems likely that this process will result in a considerable development of new guidelines, aimed at ensuring that multinationals are not using modern tax management strategies to avoid paying very much in taxes anywhere.[50]

---

43 See www.oecd.org/daf/inv/corporateresponsibility/43175359.pdf.

44 www.oecd.org/industry/inv/corporateresponsibility/43175359.pdf). There are still considerable problems at this site, involving the displaced community of Tabaco, which Cerrejón and the Colombian public security forces are said to have violently expropriated in 2001. The Independent Panel told Cerrejón to resettle Tabaco, as did the Colombian Supreme Court. The company purchased some land for this purpose, but turned over the responsibility for building and running the resettlement to the Hato Nuevo municipality, which appears to lack the capacity to manage this complex process. Last year, the company finally signed the land over to Tabaco. But 17 years after being expropriated and 10 years after the independent panel review and OECD case, Tabaco is still a displaced community. Author communication with Professor Emma Banks, 21 August 2018.

45 Canadian National Contact Point for the OECD Guidelines for Multinational Enterprises, 3 May 2011, www.oecd.org/daf/inv/mne/48754883.pdf.

46 Gold Mining in Mongolia, OECD: Responsible Business Conduct: OECD Guidelines for Multinational Enterprises, 3 May 2011, http://mneguidelines.oecd.org/database/instances/ca0009.htm.

47 See, for example, http://mneguidelines.oecd.org/database/instances/ca0009.htm.

48 OECD (2017), *OECD Transfer Pricing Guidelines for Multinational Enterprises and Tax Administrations 2017* (OECD Publishing, Paris), http://dx.doi.org/10.1787/tpg-2017-en.

49 See "Base Erosion and Profit Shifting", OECD: Better Policies for Better Lives, www.oecd.org/tax/beps/. See also Intergovernmental Forum on Mining Minerals Metals and Sustainable Development, "Base Erosion and Profit Shifting", www.igfmining.org/guidance-for-governments/base-erosion-and-profit-shifting/.

50 "Apple said ... that it had accumulated $252 billion of cash, 94% of it held offshore. That is largely a result of the tax avoidance strategies that shift profits from sales in Europe, Asia, and the Middle East out of any country's direct tax jurisdiction." "How Apple Uses the Channel Island of Jersey in Tax Strategy", *Fortune*, 6 November 2017, http://fortune.com/2017/11/06/apple-tax-avoidance-jersey/.

Over recent years there has been growing concern about transfer pricing abuse, and the difficulty developing countries face administering the arm's length principle. But, instead of getting simpler, international transfer pricing guidance gets more complex each year. The "big fixes" to overhaul the international tax system, namely formulary apportionment, and the destination based cash flow tax, face significant political obstacles, and are unlikely to deliver benefits for resource-rich developing countries. However, it is possible that in the extractive sector, solutions that require less international coordination and political goodwill, lie closer to home.[51]

These issues are important to all economic sectors. But they may be especially critical for the mining industry, which does business in so many places where tax authorities have very limited capacity and where governments have critical needs for revenue.

2.3 International Finance Corporation Performance Standards

The World Bank Group is one of the Bretton Woods institutions created at the end of World War II. Its initial role was to provide assistance for the economies damaged by World War II; indeed, the name of the initial World Bank institution was the International Bank for Reconstruction and Development (IBRD).[52] It was accompanied by the International Development Association (IDA),[53] which helps countries that do not in general have the ability to repay loans.

The World Bank has undergone a number of changes of focus and mission. The first occurred when, in the late 1950s, it was clear that the war-torn economies of the 1940s were largely recovering. This coincided with the great wave of independence and decolonisation, in which long lists of former French, British, Dutch and Portuguese colonies became independent states.[54] These countries had a long list of accumulated economic and social problems, which colonial regimes had mostly failed to solve, and a need for development assistance. The IBRD and IDA focused much of their attention on these countries.

Another great policy shift occurred in the 1980s and coincided with the Reagan government in the United States, the Thatcher government in the United Kingdom and other like-minded governments. This change in view is often referred to as the 'Washington Consensus'.[55]

---

51 Alexandra Redhead, Centre for Global Development: "What Mining Can Learn From Oil: a study in of special transfer pricing practices in the oil sector, and their potential application to hard rock minerals" (2018).

52 "Created in 1944 to help Europe rebuild after World War II", the IBRD operates by lending money to governments for development projects. See World Bank, "International Bank for Reconstruction and Development", www.worldbank.org/en/who-we-are/ibrd.

53 See International Development Association, http://ida.worldbank.org/.

54 The African countries that became independent in this period are listed by MyAfrica Now at www.myafricanow.com/african-countries-independence-days-dates/. Of course, they had many counterparts in Asia and elsewhere.

55 See "Washington consensus – definition and criticism", *Economics Help: Helping to Simplify Economics*, www.economicshelp.org/blog/7387/economics/washington-consensus-definition-and-criticism/.

Under the Washington Consensus policies, the role of government was not to execute state-led development projects, but to create frameworks of law and policy that would be attractive to private trade and investment. Private investment flows, rather than government-financed projects, were to lead economic and social development.

This implied shifting much of the money and the policy 'action' in the World Bank Group to different institutions: the International Finance Corporation (IFC),[56] which lends money to private companies to encourage their investment in developing countries; and the Multilateral Investment Guarantee Agency (MIGA),[57] which guarantees private loans to companies investing in the developing world.

In the meantime, however, the IBRD had encountered a series of projects that triggered strong adverse public reactions – notably a series of dam projects which flooded river valleys, displacing large numbers of poor farmers, who were supposed to be the kinds of people that development agencies were trying to help.[58] This led the IBRD to develop a series of operational policies or safeguards designed to provide some 'rules of the road' in dealing with particularly fraught issues such as displacement and relocation of populations, a requirement for environmental assessment of all large projects, projects in traditional indigenous territories, projects that impacted on natural habitats and so on.[59]

The growing role of the IFC and MIGA therefore created a challenge: were these policies to apply to loans to the private sector?

For some time, the IFC equivocated and temporised. Finally, a particularly controversial and difficult hydroelectric project in Chile, the Pangue Dam, raised the issue to a level that provoked the president of the World Bank Group

---

56 See IFC, "Overview", www.ifc.org/wps/wcm/connect/corp_ext_content/ifc_external_corporate_site/about+ifc_new.

57 See Multilateral Investment Guarantee Agency, www.miga.org/.

58 "The Sardar Sarovar projects on the Narmada River in western India, and the World Bank's role in supporting them until 1993, have sparked worldwide controversy." Independent Evaluation Group, http://lnweb90.worldbank.org/oed/oeddoclib.nsf/DocUNIDViewForJavaSearch/12A795722EA20F6E852567F5005D8933. See also Nayak, "Big Dams and Protests in India: A Study of Hirakud Dam", *Economic and Political Weekly* (2010); Mekong Commons, "Pak Mun Dam: 25 years after World Bank's loan, problems remain" (2016), www.mekongcommons.org/pak-mun-dam-25-years-world-banks-loan-problems-remains/; "When the World Bank Does More Harm Than Good", NPR.com (17 April 2015), www.npr.org/sections/goatsandsoda/2015/04/17/399816448/when-the-world-bank-does-more-harm-than-good; "Are Failed Infastructure Projects Linked to the Presence of the IMF or World Bank?" Knowledge at Wharton (7 September 2005), http://knowledge.wharton.upenn.edu/article/are-failed-infrastructure-projects-linked-to-the-presence-of-the-imf-or-world-bank-read-on/; Sasha Chavkin *et al*, "How the World Bank Broke its Promise to Protect the Poor", *The Huffington Post* (2015), http://projects.huffingtonpost.com/projects/worldbank-evicted-abandoned.

59 World Development Report 1989. Oxford University Press https://openknowledge.worldbank.org/bitstream/handle/10986/5972/WDR%201989%20-%20English.pdf?sequence=1; Devesh Kapur *et al*, *The World Bank: Its First Half Century* (Washington DC: Brookings Institution Press, 1997). https://books.google.com/books?id=mtBFK_ACbv8C&pg=PA1028&lpg=PA1028&dq=IBRD+operational+policies+1980s&source=bl&ots=3Fyb8Simsf&sig=mYkxmBOlBxfKc_npzgoECefsqJ8&hl=es&sa=X&ved=2ahUKEwiNiOnshcrcAhX0OX0KHabgAw4Q6AEwCHoECAgQAQ#v=onepage&q&f=false Review of World Bank Conditionality: Operations Policy and Country Services, World Bank, September 2005; http://siteresources.worldbank.org/PROJECTS/Resources/40940-1114615847489/webConditionalitysept05.pdf.

to appoint an independent review commission. Its report,[60] which found substantial non-compliance by the IFC, and the subsequent debate within the IFC led to a clear decision that such policies would be applicable to private sector loans. This led to the establishment of the initial IFC Performance Standards, the first version of which was published in 1998.[61]

These imposed requirements on borrowers in the areas set out in Table 1.

Table 1. IFC Performance Standards 1998

Environmental assessment	International waterways
Natural habitats	Indigenous peoples
Pest management	Involuntary resettlement
Forestry	Cultural property
Safety of dams	Child and forced labour

This set of standards was revised in 2006[62] and again in 2012.[63] It has become by far the most widely recognised and applied set of environmental and social performance standards for private foreign investment in developing countries. Companies that cannot or will not comply with these standards are not eligible for IFC loans or MIGA guarantees.

Further, when it decided to make these standards applicable to its private sector loans, the IFC set up a monitoring system and recourse mechanism under an independent office, the Compliance Adviser/Ombudsman (CAO).[64]

People impacted by IFC financed projects can file complaints with the CAO, which will then investigate and issue reports. At this point, the CAO has investigated a considerable number of cases, many of them in the mining industry.[65] The work of the CAO in supporting better performance and in helping companies to build their capacity to manage projects better has had a strong and positive impact on company performance.

60 Jay D Hair, L Danielson and B Dysart, *Pangue Hydroelectric Project (Chile): An Independent Review of the International Finance Corporation's Compliance with Applicable World Bank Group Environmental and Social Requirements* (14 April 1997), www.sdsg.org/wp-content/uploads/2010/02/Pangue-hydroelectric-project.pdf.

61 See Safeguard Policies, World Bank Group, www.ifc.org/wps/wcm/connect/topics_ext_content/ifc_external_corporate_site/sustainability-at-ifc/policies-standards/safeguards-pre2006#safeguard.

62 IFC Sustainability Framework, 2006 Edition, www.ifc.org/wps/wcm/connect/topics_ext_content/ifc_external_corporate_site/sustainability-at-ifc/policies-standards/ifcsustainabilityframework_2006.

63 www.ifc.org/wps/wcm/connect/topics_ext_content/ifc_external_corporate_site/sustainability-at-ifc/policies-standards/ifcsustainabilityframework_2012.

64 www.cao-ombudsman.org/.

65 www.cao-ombudsman.org/cases/.

The impact has been greatly enhanced by the development of the Equator Principles. In general, the IFC has operated by participating in and leading loan syndications. Typically, the IFC will make only a minority percentage of loans, but lead the group of lenders. The remainder is typically lent by private banking institutions. This has meant that these private banks have gained considerable experience working with the IFC and its standards. And as they themselves came under pressure from civil society critics, they tended to adopt similar approaches.

These have been formalised in a series of agreements among private financial institutions called the Equator Principles.[66] Under this system, 94 private financial institutions in 37 countries have adopted the Equator Principles.[67] The participants in the system agree to a number of measures. The most salient is their agreement to adopt and apply the IFC standards in making loans.[68]

Thus, the standards developed by the IFC have to a large degree become the common currency among large lenders to the private sector. Similar sets of standards have also been developed by regional development banks such as the Inter-American Development Bank,[69] the Asian Development Bank,[70] the African Development Bank[71] and the European Bank for Reconstruction and Development.[72]

The New Development Bank[73] – an institution supported by the BRICS countries as an alternative to the World Bank – has developed a system of rules that is similar in concept.[74] The Asia Infrastructure Investment Bank, backed by the Chinese government,[75] also has an environmental and social sustainability framework.[76] What may be missing is a system equivalent to the Equator Principles for private capital sources located in emerging economies such as China, India, Russia and Brazil.

While it has its share of problems and limitations, this system of enforcement of social and environmental performance by lending institutions has had considerable impact.

66 http://equator-principles.com/.

67 http://equator-principles.com/members-reporting/.

68 Emir Basic, "The Equator Principles: An Empirical Study of Their Effect on the Financial Performance of Banks", pp5-6, www.researchgate.net/publication/296486322._

69 www.iadb.org/en/about-us/sustainability-and-safeguards.

70 www.adb.org/site/safeguards/main.

71 www.afdb.org/en/news-and-events/afdb-launches-revised-version-of-its-environmental-and-social-assessment-procedures-for-2015-15013/; /www.afdb.org/en/projects-and-operations/procurement/resources-for-borrowers/policies-procedures/.

72 www.ebrd.com/news/publications/policies/environmental-and-social-policy-esp.html.

73 www.ndb.int/.

74 www.inesc.org.br/news/2017/april/analysis-of-the-brics-new-development-bank-environment-and-social-framework.

75 "China launches new AIIB development bank as power balance shifts", Reuters, 15 January 2016, www.reuters.com/article/us-asia-aiib-investment/china-launches-new-aiib-development-bank-as-power-balance-shifts-idUSKCN0UU03Y.

76 www.aiib.org/en/policies-strategies/framework-agreements/environmental-social-framework.html.

3. Treaties

Treaties are the most obvious limitation on principles of national sovereignty. While typically treaties do not act directly on private participants in trade and investment markets, they do create a common platform for participating states.[77] And the laws and rules of these states can and do have a direct impact on private participants in the sector.

While few, if any, treaties are aimed specifically at the mining industry, a considerable number have an impact on the industry in one way or another.

3.1 Transnational space

One widely discussed way in which mining is affected by treaties is the treaty regime applicable to so-called 'transnational space'. These are areas that are not subject to the control and management of sovereign national governments.

They include the high seas outside areas of national control. The United Nations Convention on the Law of the Sea (1982)[78] (UNCLOS) provides that minerals found in the area are "the common heritage of mankind",[79] and cannot be exploited except through compliance with "the rules, regulations and procedures" developed by the International Seabed Authority, an agency established by the treaty.[80]

The United States has not ratified UNCLOS and asserts a right to act unilaterally: "It is the legal position of the United States that exploration for and commercial recovery of hard mineral resources of the deep seabed are freedoms of the high seas subject to a duty of reasonable regard to the interests of other states in their exercise of these and other freedoms recognized by general principles of international law."[81]

A second area of transnational space managed through treaty arrangements is the Antarctic.[82] The original Antarctic Treaty did not mention mining. But mining was later explicitly prohibited in:

> *the Protocol on Environmental Protection to the Antarctic Treaty (an addition to the treaty), it is also known as the Madrid Protocol. It was signed in 1991 in Madrid, Spain by the signatories to the Antarctic Treaty banning mining. The protocol entered into force (became law) on the 14th of January 1998. It is up for review after 50 years in 2048 when any changes will need at least a 75% vote from the consultative parties.*[83]

---

77 They may eventually be seen as establishing a standard of customary international law that could also become binding on non-signatory states. George K Walker, "Sources of International Law and the Restatement (Third), Foreign Relations Law of the United States", 37 Naval L 1, 10-12 (1988).

78 United Nations website, www.un.org/depts/los/convention_agreements/convention_overview_convention.htm.

79 United Nations Convention on the Law of the Sea, Article 136, opened for signature 10 December 1982, 1833 UNTS 397 (entered into force 16 November 1994).

80 *Id* at Article 137(2).

81 30 USC § 1401(12).

82 See the Antarctic Treaty, Secretariat of the Antarctic Treaty, www.ats.aq/e/ats.htm.

83 "Human Impacts on Antarctica and Threats to the Environment – Mining and Oil", *Cool Antarctica*, www.coolantarctica.com/Antarctica%20fact%20file/science/threats_mining_oil.php.

The commitment of the Madrid Protocol has been more recently reaffirmed.[84]

Finally, as technological advances make commercial activities on the moon or elsewhere in space more imaginable, there is a treaty regime in place.[85] The treaty system is complex and not free from doubts. Among other things, there is a debate over the extent to which the Outer Space Treaty applies to any exploitation of asteroids: "In 2015, the United States passed the U.S. Commercial Space Launch Competitiveness Act, which in a nutshell allows for U.S. citizens to exploit asteroids and other space resources, but not the land on which the resources sit. While this makes resource hunting legal for U.S. citizens, some experts have said this could violate the Outer Space Treaty."[86]

Some sense of where this discussion may be going can be gleaned from the guidelines developed to this stage by the Space Resources Governance Working Group.[87]

3.2 Labour conditions

The International Labour Organisation (ILO) dates to 1919.[88] The ILO was one of the few pieces of the League of Nations system to survive the collapse of the league. Its function has been to ensure the protection of workers and the improvement of working conditions.[89] Relations with mine workers, their unions and the legal context of mining employment have been a major issue in the industry throughout its history. And much of the progress to date has come through the development and promotion of treaties.

The basic rights of workers, including the right to organise and form unions, have been recognised by the United Nations in Articles 23 and 24 the Universal Declaration of Human Rights.[90] But ILO-initiated treaty processes have put some considerable meat on these bones.

There are conventions and agreements for the prohibition of forced labour and child labour, practices that persist in the darker corners of the industry. While the reputable companies are not engaging in these practices, it is difficult to eliminate them from corporate supply chains:

---

84 "The 29 countries party to the Antarctic Treaty unanimously reaffirmed their commitment to a ban on mining activities in the Antarctic" in 2016. "Antarctic Mining Ban Reaffirmed", *Maritime Executive*, www.maritime-executive.com/article/antarctic-mining-ban-reaffirmed#gs.dBGUb4Y.

85 Treaty Governing the Activities of States in the Exploration and Use of Outer Space, Including the Moon and Celestial Bodies, opened for signatures 27 January 1967, 18 UST 2410, TIAS 6347, 610 UNTS 205 (entered into force 10 October 1967).

86 Elizabeth Howell, "Who Owns the Moon? | Space Law & Outer Space Treaties", *Space* (27 October 2017), www.space.com/33440-space-law.html. The United States is not the only country to move in this direction. See Grand Duchy of Luxembourg, "On the Exploration and Use of Space Resources", 1 August 2017.

87 See https://swfound.org/news/all-news/2017/09/the-hague-space-resources-governance-working-group-releases-draft-building-blocks.

88 ILO, "About the ILO", www.ilo.org/global/about-the-ilo/lang—en/index.htm.

89 ILO, "Mission and Impact of the ILO", www.ilo.org/global/about-the-ilo/mission-and-objectives/lang—en/index.htm

90 https://treaties.un.org/doc/Treaties/1990/12/19901218%2008-12%20AM/Ch_IV_13p.pdf.

The prohibition of slavery, forced labour, institutions and practices similar to slavery and trafficking in persons are enshrined in international law, however debates about the relationship between these concepts – and how they should be translated into domestic law – have at times led to confusion about how best to tackle the enormous challenges posed by contemporary forms of forced labour or what is often called "modern forms of slavery".

***

In June 2014, the ILO adopted a new legally binding Protocol to the Forced Labour Convention, 1930, supported by a Recommendation on Supplementary Measures for the Effective Suppression of Forced Labour (No. 203), to strengthen global efforts to eliminate forced labour. The new instruments bring the ILO's standards against forced labour into the twenty-first century to address effectively all forms of modern-day forced labour, including human trafficking.[91]

And there is much recent attention to child labour. Convention 182 on the Worst Forms of Child Labour was adopted in June 1999.[92] Child labour in mining is not a thing of the past:

Gold mining is extremely dangerous work for children. Yet still today, tens of thousands are found in the small-scale gold mines of Africa, Asia and South America. Children work both above and under ground. In the tunnels and mineshafts they risk death from explosions, rock falls, and tunnel collapse. They breathe air filled with dust and sometimes toxic gases. Above ground, children dig, crush, mill, and haul ore – often in the hot sun.[93]

Major companies, including major mining companies, generally do not employ children directly or use forced labour in their own operations. They do, however, have a hard time keeping these practices out of their supply chains.[94] "A report by Amnesty International first revealed that cobalt mined by children was ending up in products from several companies, including Apple, Microsoft, Tesla and Samsung."[95]

The difficulty that many companies have faced in trying to comply with the conflict minerals supply chain due diligence requirements of the US Dodd Frank conflict minerals rules[96] is illustrative of the challenge of keeping questionable

91 Lee Swepston, "Forced and compulsory labour in international human rights law", ILO Working Paper, www.ilo.org/global/topics/forced-labour/publications/WCMS_342966/lang—en/index.htm. Paper presented at the conference, "Shaping the Definition of Human Trafficking", Dickson Poon School of Law, King's College London, May 2014, www.ilo.org/wcmsp5/groups/public/—-ed_norm/—-declaration/documents/publication/wcms_342966.pdf.

92 Convention 182, www.ilo.org/ipec/Campaignandadvocacy/Youthinaction/C182-Youth-orientated/C182Youth_Convention/lang—en/index.htm.

93 ILO, "Learn More About Child Labour in Mining", www.ilo.org/ipec/areas/Miningandquarrying/MoreaboutCLinmining/lang—en/index.htm.

94 See, for example, "Modern slavery, forced labour and human rights: The business case beyond compliance", Innovation Forum, https://innovationforum.co.uk/conferences/modern-slavery-human-rights/.

95 "CBS News finds children mining cobalt for batteries in the Congo", *CBS This Morning*, 5 March 2018, www.cbsnews.com/news/cobalt-children-mining-democratic-republic-congo-cbs-news-investigation/.

96 See note 8, *supra*. The US administration seems intent on withdrawing these rules.

labour practices out of the mineral supply chain; as is the complexity of the Kimberley Process,[97] designed to keep 'blood diamonds', many of them produced by children, out of the global supply chain.[98]

3.3 Environmental agreements

There is a plethora of environmental agreements that can in some circumstances be relevant to the mining and minerals industries. Among the more prominent is the Basel Convention,[99] which generally prohibits the transboundary shipment of hazardous waste and has on occasion blocked the shipment of waste containing recoverable minerals.[100] The Convention on Biological Diversity[101] and the Minimata Convention on Mercury[102] are also important for the minerals industries.

A working group of scientists has attempted to define the ecological limits beyond which the Earth could become uninhabitable: the nine 'planetary boundaries'.[103] It is hard to see most of these global environmental threats being successfully dealt with without some kind of global agreement on managing them and hard to see how most of them can be resolved without the minerals industries doing their appropriate share.

3.4 Rights of indigenous peoples

In many parts of the world, there is a high level of indigenous activism and a very rapidly changing legal framework for development of mineral resources on traditional indigenous lands, or which affect indigenous communities or resources.

Much of the legal foundation for this dynamic set of developments is ILO Convention 169 (1982), the Indigenous and Tribal Peoples Convention.[104] The

97 "The Kimberley Process started when Southern African diamond-producing states met in Kimberley, South Africa, in May 2000, to discuss ways to stop the trade in 'conflict diamonds' and ensure that diamond purchases were not financing violence by rebel movements and their allies seeking to undermine legitimate governments." Kimberley Process, www.kimberleyprocess.com/en/about.

98 Global Witness, "Conflict Diamonds", www.globalwitness.org/en/campaigns/conflict-diamonds/.

99 The Basel Convention on the Control of Transboundary Movements of Hazardous Wastes and Their Disposal, S Treaty Doc 5 (1991), 28 ILM 657, available in United Nations Environment Programme, Basel Convention on the Control of Transboundary Movements of Hazardous Wastes and Their Disposal www.basel.int/TheConvention/Overview/TextoftheConvention/tabid/1275/Default.aspx.

100 *Id* at 681.

101 Convention on Biological Diversity, 5 June 1992, 1760 UNTS 143, www.cbd.int/doc/publications/cbd-sustain-en.pdf.

102 See Office of High Commissioner, www.ohchr.org/en/issues/ipeoples/pages/declaration.aspx; www.mercuryconvention.org/Portals/11/documents/conventionText/Minamata%20Convention%20on%20Mercury_e.pdf.

103 The 'planetary boundaries' are, in one formulation, ocean acidification, climate change, overabundance of phosphorous and nitrogen, species extinction, atmospheric aerosol loading, fresh water supplies, land sue change, stratospheric ozone depletion and chemical pollution. Rockstrom, "A safe operating space for humanity", *Nature* (September 2009), www.fs.fed.us/pnw/pubs/journals/pnw_2009_rockstrom001.pdf. See also "The nine planetary boundaries", Stockholm Resilience Centre, www.stockholmresilience.org/research/planetary-boundaries/planetary-boundaries/about-the-research/the-nine-planetary-boundaries.html.

104 www.ilo.org/dyn/normlex/en/f?p=NORMLEXPUB:12100:0::NO::P12100_ILO_CODE:C169.

principles of this convention have been further elaborated in the 'soft law' United Nations Declaration on the Rights of Indigenous Peoples (2007).

The convention has had a considerable effect on international law. In the leading case of *Kichwa Indigenous People of Sarayaku v Ecuador*,[105] involving oil and gas exploration in indigenous territory, the Inter-American Court of Human Rights determined that Ecuador had failed in its duty to protect the rights of the indigenous peoples by failing to seek their free prior informed consent to the activities conducted by the oil company.

It has also had an important impact on the domestic law of many countries that are signatories to the convention. Courts in Chile, Colombia, Peru, Canada and elsewhere have recognised the convention as establishing various rights and remedies under domestic law. Many of these cases are related to mining and to mineral exploration. The Ecuadorian courts recently invalidated a grant of mineral concessions for (among other reasons) failure to consult adequately with indigenous communities.[106]

3.5 Investor protection and dispute resolution

Perhaps no development in recent years is as relevant to the relationship between host states and investors as the advent of large-dollar arbitration claims against governments that are said to have disappointed the expectations of investors – whether the claim of Crystallex against Venezuela,[107] the Pacific Rim claims against El Salvador,[108] Bear Creek Mining's claims against Peru,[109] several claims against Colombia[110] or the recent claim of Gabriel Resources against Romania.[111]

The potential that a disappointed investor may be able to obtain a multibillion-dollar award fundamentally shifts the balance of power between government and foreign investor. The fact that the foreign investor may be able to obtain third-party funding to pursue its claims in these immensely expensive legal processes further tips the balance.

These developments are much facilitated by the adoption of the Convention on the Settlement of Investment Disputes between States and Nationals of

105 *Kichwa Indigenous People of Sarayaku v Ecuador*, Merits and Reparations, Inter-Am. Ct HR (ser C) No 245 (27 June 2012).

106 *Corte Provincial del Azuay, Juicio* 01333201803145, 3 August 2018.

107 *Crystallex International Corporation v Bolivarian Republic of Venezuela*, ICSID Case ARB(AF)/11/2, 4 April 2016.

108 Claimant's Notice of Arbitration at 53, *Pac Rim Cayman LLC v Republic of El Salvador*, ICSID Case ARB/09/12 (registered 15 June 2009), available at www.pacrim-mining.com/i/pdf/2009-04-30_CAFTAF.pdf.

109 See Daniela Páez-Salgado, "Four Key Takeaways of the Decision in *Bear Creek Mining Corp v Republic of Peru*", Kluwer Arbitration Blog, 16 December 2017, http://arbitrationblog.kluwerarbitration.com/2017/12/16/bear-creek-mining-v-peru.

110 "Three mining disputes: the first investment disputes against Colombia come to light", *Investment Treaty News* (2016), www.iisd.org/itn/2016/05/16/three-mining-disputes-the-first-investment-disputes-against-colombia-come-to-light/.

111 Gabriel Resources LTD and Gabriel Resources (Jersey) Romania, International Arbitration Database, https://icsid.worldbank.org/en/Pages/cases/casedetail.aspx?CaseNo=ARB/15/31.

Other States (1966) (the ICSID Convention)[112] and the Convention on the Recognition and Enforcement of Foreign Arbitral Awards (1958) (the New York Convention).[113] While a right of arbitration does not always depend upon these conventions (the right may come from bilateral investment treaties or by contract between host country and investor), the conventions have been a major factor in the increased frequency and seriousness of claims, in a system where government cannot ever really 'win'.

There is a real concern that governments will be dissuaded by threats of arbitration from enforcing environmental laws, protecting land rights or otherwise balancing the rights of its citizens against the commercial rights of investors.

4. Voluntary standards and rules

The governance vacuum in the minerals sector at the international level has been so considerable that various combinations of private sector, government and civil society actors have devised a variety of systems of voluntary standards to help fill the void.[114] Clearly, systems that are voluntary and do not carry the weight of government sanctions behind them have some limitations. Just as clearly, some of these systems have had important and positive effects.

They can certainly be adopted more quickly and through less cumbersome processes than treaty negotiation, can continue to evolve more rapidly than treaty regimes and can probably enter into greater levels of detail more quickly.

4.1 The turn of the century 'explosion' of standards

In 1998 the first full set of the IFC Performance Standards was established.[115] Around that time, the intense pressure on industry to improve its performance led to a series of important and influential voluntary initiatives. The reasons for this explosion of initiatives have been explored by a number of authors.[116]

(a) Global Reporting Initiative

The Global Reporting Initiative (GRI) is an independent international organisation that helps businesses, governments and other organisations to understand and communicate the impact of business on critical sustainability issues such as climate change, human rights, corruption and many others.[117] It sets reporting standards for sustainability reporting by corporations.

112 Settlement of Investment Disputes, 27 August 1965, 17 UST 1270.

113 Convention on the Recognition and Enforcement of Foreign Arbitration, 29 December 1970, 21 UST 2517.

114 See L Danielson, "Evolution of Global Standards in the Extractive Sector: the Growing Role of the World Bank Group", Chapter 1.7 in Bastida *et al* (eds), *International and Comparative Mineral Law and Policy: Trends and Prospects* (Kuler 2005).

115 See note 61, *supra*.

116 See note 20, *supra*.

117 www.globalreporting.org/Pages/default.aspx.

While the GRI applies to industry broadly, and not just the mining industry, mining companies were among its early and influential supporters. They have continued to lead in the creation of an industry-specific mining and metals sector supplement, with specific reporting standards tailored to the minerals industries, which is now in its fourth version.[118] Most major mining companies report publicly on their environmental and social performance under these guidelines.

(b) *Mining, Minerals and Sustainable Development Project*

In 1999 mining industry leaders, in consultation with the International Institute for Environment and Development, decided to launch a Global Mining Initiative, one element of which would be a multi-year process of stakeholder consultation and research headquartered at the International Institute for Environment and Development. The resulting research project was the largest research project ever undertaken on the environmental, economic and social impacts of the minerals industries.[119]

The Mining, Minerals and Sustainable Development Project (MMSD) conducted a wide variety of research on multiple topics,[120] established several regional processes around the world that examined these issues regionally[121] and issued a final report,[122] whose recommendations formed the work plan of the newly formed International Council on Mining and Metals, an industry organisation. The final MMSD report was submitted to the 2002 World Summit on Sustainable Development in Johannesburg.[123] The MMSD recommendations influenced the International Council on Mining and Metals (ICMM)) Principles, the 2006 IFC Performance Standards, the Intergovernmental Forum and a number of the other initiatives discussed in this chapter.

(c) *ICMM Principles*

The ICMM is a global industry association which brings together 27 mining and metals companies as well as over 30 national and regional mining associations and global commodity associations to maximise the contribution of mining, minerals and metals to sustainable development.[124] In 2003 the ICMM developed a set of 10 principles to which its members must subscribe;[125] it has also developed a system of assurance.

---

118 www.globalreporting.org/resourcelibrary/GRI-G4-Mining-and-Metals-Sector-Disclosures.pdf.
119 www.iied.org/mining-minerals-sustainable-development-mmsd.
120 www.iied.org/mmsd-working-papers
121 www.iied.org/mmsd-regional-reports.
122 *Breaking New Ground* (2002), www.iied.org/mmsd-final-report.
123 See Global Public Policy Institute, "Architecture for Change: An Account of the Mining, Minerals and Sustainable Development Project" (2012); A Azapagic, "Developing a framework for sustainable development indicators for the mining and minerals industry", (2004).
124 www.icmm.com/en-gb/members.
125 www.icmm.com/en-gb/about-us/member-commitments/icmm-10-principles.

(d) *Extractive Industries Transparency Initiative*

The Extractive Industries Transparency Initiative (EITI) seeks to address the issues of corruption and transparency in the mining and oil and gas industries. It was established in 2002[126] to promote and develop a global standard for the open and accountable management of extractive resources, and is overseen by multi-stakeholder boards at both the global and national levels. The EITI applies a set of standards last updated in 2016.[127]

It requires reporting of payments by companies to government and government officials at all levels, and reconciliation of reports from companies and participating governments. National compliance with EITI requirements is independently validated. Most observers seem to believe it has had some real achievements in combating corruption. The next step may be to move beyond transparency of revenue flows to promote transparency of the underlying agreements on which revenues are supposed to be based.

(e) *Voluntary Principles on Security and Human Rights*

The Voluntary Principles on Security and Human Rights,[128] established in 2000, are a set of principles, aimed specifically at the extractive industries, designed to guide companies in maintaining the safety and security of their operations within an operating framework that encourages respect for human rights. In many countries, national police and military forces have a record of human rights abuse. Company security forces naturally need to work with these government institutions, but the voluntary principles attempt to define the kinds of engagement that are – and are not – consistent with corporate respect for human rights. And the organisation provides a variety of guidance tools on implementing the principles.[129]

(f) *International Cyanide Management Code for the Manufacture, Transport, and Use of Cyanide in the Production of Gold*

The Cyanide Code,[130] established in 2003, was developed by a multi-stakeholder steering committee under the guidance of the United Nations Environmental Programme and the then-International Council on Metals and the Environment, whose functions have been subsumed by the ICMM. It has an auditing process and a dispute resolution system.

(g) *Responsible Jewellery Council*

The Responsible Jewellery Council[131] is an international not-for-profit

---

126 https://eiti.org/history.
127 https://eiti.org/document/standard.
128 www.voluntaryprinciples.org/.
129 www.voluntaryprinciples.org/files/Implementation_Guidance_Tools.pdf.
130 International Cyanide Management Code: For the Gold Mining Industry, www.cyanidecode.org/.
131 Responsible Jewellery Council, www.responsiblejewellery.com/.

organisation established in 2005, which sets standards and performs certification. It brings together more than 450 member companies and is committed to promoting responsible ethical, human rights, social and environmental practices in a transparent and accountable manner throughout the jewellery industry from mine to retail

(h) Intergovernmental Forum on Mining Metals Minerals and Sustainable Development

The Intergovernmental Forum (IGF)[132] is different from the previous initiatives in that it is aimed not at improving private sector performance, but at strengthening appropriate government management of the sector. It was founded at the Johannesburg Earth Summit in 2002. It now has some 67 member governments.

The IGF has a Mining Policy Framework,[133] which is updated and improved from time to time, and consists of a suite of recommended policy elements which are designed to promote laws, policies and institutions necessary for the support of a minerals sector that contributes more fully to sustainable development. The IGF offers member countries the opportunity to have their legal and policy framework assessed by expert teams against this framework,[134] and provides a range of capacity building services.

4.2 More recent development of systems

Since the proliferation of standards in the years around 2000, the evolution and elaboration of additional sets of norms have continued.

(a) International Organisation for Standardisation

The International Organisation for Standardisation (ISO) has developed a wide range of technical standards used in the mining industry.[135] More recently, ISO has begun to provide standards for environmental and social management, the best known of which are probably the environmental standards (ISO 14000) and the quality management standard (ISO 9001).

ISO 14000 family – Environmental management:

The ISO 14000 family of standards provides practical tools for companies and organizations of all kinds looking to manage their environmental responsibilities. ISO 14001:2015 and its supporting standards such as ISO 14006:2011 focus on environmental systems to achieve this. The other standards in the family focus on

132 Intergovernmental Forum on Mining, Minerals, Metals, and Sustainable Development, http://igfmining.org/.

133 Intergovernmental Forum on Mining, Minerals, Metals, and Sustainable Development: Mining Policy Framework, http://igfmining.org/mining-policy-framework/.

134 Intergovernmental Forum on Mining, Minerals, Metals, and Sustainable Development: Assessments and Reports, http://igfmining.org/mpf-reports/.

135 See, for example, the standards listed at www.iso.org/ics/73/x/.

specific approaches such as audits, communications, labelling and life cycle analysis, as well as environmental challenges such as climate change.[136]

ISO has also developed guidance for corporate social responsibility:

ISO 26000:2010 provides guidance rather than requirements, so it cannot be certified to unlike some other well-known ISO standards. Instead, it helps clarify what social responsibility is, helps businesses and organizations translate principles into effective actions and shares best practices relating to social responsibility, globally. It is aimed at all types of organizations regardless of their activity, size or location.[137]

ISO appears to be using this as a platform for the development of related social management guidance, such as human rights. This has not been without controversy.[138]

There is a new ISO sub-committee on mining reclamation management. which intends to develop international standards for managing mine reclamation.[139]

(b) Natural Resource Charter

The Natural Resource Charter is an independent initiative intended to complement the EITI in spelling out the entire decision chain "by which natural assets can become a blessing instead of a curse".[140] It is focused on improved government management of mineral revenues. Developed by leading economists and governance experts, it consists of a series of 12 principles, directed to governments, for improved management of the minerals sector. There is a system of some 170 benchmarking questions to be used in evaluating progress in reaching the charter's goals.[141]

In its origin, it focused mostly on the relationship between resource investors and national government, but it has now begun to work on issues of sub-national governance and revenue sharing.

(c) Ruggie Principles

The United Nations Human Rights Council designated Harvard Professor John Ruggie as a special rapporteur on business and human rights. In 2011 he produced the United Nations Guiding Principles on Business and Human Rights.[142] These are

---

136 www.iso.org/iso-14001-environmental-management.html.

137 www.iso.org/iso-26000-social-responsibility.html.

138 The chair of the UK committee on ISO 26000, Adrian Henriques, has written: "The development of ISO 26000 has been fraught with politics, yet it is a powerful statement of the moral and practical issues facing companies and other organisations." But see 31 August 2017 letter to Technical Committee of ISO from a group of non-governmental organisations (NGOs), https://static1.squarespace.com/static/583f3fca725e25fcd45aa446/t/59ae0c119f7456bae983ed04/1504578599102/Letter+to+ISO+opposing+Technical+Committee+for+Social+Responsibility.FINAL.pdf.

139 www.iso.org/news/2015/02/Ref1933.html.

140 https://resourcegovernance.org/approach/natural-resource-charter.

141 www.opengovpartnership.org/stories/natural-resource-charter-benchmarking-framework-170-crucial-questions-resource-rich.

a set of guidelines for states and companies to prevent, address and remedy human rights abuses committed in business operations. The principles have generally been well received in the mining industry and are increasingly influential. A number of companies have taken steps to internalise them in management.

(d) Responsible Mining Index

The Responsible Mining Index, published by the Responsible Mining Foundation,[143] "is the first ranking created exclusively for the mining sector. The RMI has ranked the 30 biggest mining companies in the world, both state-owned and publicly listed, on their performance in meeting society's expectations around environmental, social, economic, and governance issues".

The first index was published in 2018.[144]

4.3 Certifications and standards for artisanal mining

Artisanal miners may produce something in the order of 10% to 15% of the world gold supply, but constitute perhaps 90% of the world's gold mining workforce. There are approximately 20 million artisanal gold miners worldwide.[145] The artisanal gold mining industry is rife with environmental problems, use of mercury and labour abuse.

One concern is that efforts to require better standards will result in large companies complying, but will wind up excluding the small producers – which badly need the economic return – from world markets. But multiple certification systems have been developed to allow consumers to purchase artisanal gold that is certified to be free of the worst of these practices, including schemes such as Fair Mined,[146] Fair Trade[147] and CRAFT Code.[148]

The layout of these certification schemes follows a general design with some differences in detail: artisanal and small-scale mining operations bring their organisational practices into compliance with rigorous standards in social and environmental protection; those operations are then validated by independent third-party certifiers and given access to large global markets; buyers then pay a premium over the market price of gold in exchange for the heightened regulation with which the gold was extracted; and this premium is then reinvested in artisanal mining communities to pay for improving artisanal mining practices and for community development.

---

142 www.ohchr.org/Documents/Publications/GuidingPrinciplesBusinessHR_EN.pdf.

143 "The goal of the Responsible Mining Foundation is to encourage continuous improvement in responsible mining by large-scale minerals and metals mining companies (LSM) across a range of economic, environmental, social and governance (EESG)", https://responsibleminingindex.org/ru/foundation/about/who-we-are.

144 See http://rmi.vizzuality.com/en.

145 www.fairmined.org/why-fairmined/.

146 www.fairmined.org/.

147 www.fairgold.org/about-us/.

148 www.responsiblemines.org/en/our-work/standards-and-certification/craft/.

By increasing transparency in gold supply chains and decreasing information asymmetries for the consumer, these certification schemes fill the governance gap for artisanal mining operations, providing proper labour standards, environmental protections and social development programmes that would otherwise be lacking.

Obvious future steps are to move this process beyond gold to include other commodities where artisanal production is important. Silver produced as a gold byproduct is already available. But mining of gemstones – including not only diamonds, but emeralds, larimar and possibly tanzanite – may also be a candidate.

4.4 Bringing it all together – the Initiative for Responsible Mining Assurance

This brief account can hardly do justice to the energy and innovative spirit of many who have tried to improve the performance of the industry, capture more development benefits and reduce adverse environmental and social impacts. One author lists 45 such initiatives.[149] Indeed, a new gold industry system seems to have surfaced while this chapter was being written.[150]

But this discussion cannot reasonably conclude without mention of one of the newest and perhaps most ambitious of all the voluntary standards: the Initiative for Responsible Mining Assurance (IRMA).[151]

IRMA addresses many of the concerns and weaknesses of some of the other voluntary initiatives. One industry concern has been how to manage such a plethora of individual sets of requirements, from the IFC Performance Standards to all of the various voluntary standards set out above. IRMA brings together many of the essential elements of these other requirements into a single document, filling in some of the most glaring gaps in the process. Its standard[152] is a single management system that is intended to deal with the whole range of sustainability concerns.

Further, it addresses the concern that a mine may, for example, have exemplary relations with indigenous communities or outstanding environmental practices, but may still embarrass purchasers of its product by having child labour in its supply chain or human rights problems. The IRMA standard is designed to cover all the important bases, assuring buyers that the product they are purchasing is responsibly sourced.

Many criticisms of voluntary standards focus on the lack of rigorous external assurance of compliance. IRMA has as rigorous a system as there is in the industry, under the management of an independent multi-stakeholder organisation.

---

149 RA Hodge, *Towards contribution analysis*, United Nations University (January 2017).

150 Responsible Gold Mining Principles, World Gold Council, www.gold.org/who-we-are/our-members/responsible-gold/responsible-gold-mining-principles.

151 www.responsiblemining.net/.

152 www.responsiblemining.net/irma-standard.

This system is new, but has the potential to be a groundbreaking development. The incentive for getting into this system is essentially the customer base. It is no accident that one of the consistent funders of the 10-year development process was the Tiffany Foundation. Other major buyers of minerals are also very much involved in this developing system. It is likely that a growing number of customers will start asking for IRMA certification of the products they buy.

4.5 Evaluating and assessing voluntary standards

If so much of the heavy lifting of improving performance is dependent on ingeniously designed private initiatives, it is very important to work towards some kind of methodology to measure how effective they may be.[153] A simple model of the kinds of initiatives that have been developed to fill this governance vacuum would focus on three basic questions.

(a) *A standard*

First, there needs to be some sort of rule that defines the kind of conduct we seek (or prohibits certain conduct we want to discourage or eliminate). It is generally thought that the credibility and legitimacy of such rules are heavily dependent on the process by which the rules are developed. The process needs to be led by a balanced leadership group, which is transparent and open to a variety of views and interests. There is now an international organisation that certifies the balance and integrity of standards development processes: the Iseal Alliance. This idea is sometimes called 'certifying the certifiers'.[154]

The principal concerns in this area include the following:

- Vague standards: Many of the standards to date do not require anything concrete enough to tell whether someone is or is not complying (eg, "Our company seeks to do good in the communities where we operate").
- Imbalance: Some standards have been developed by industry organisations and do not adequately take into account the interests of other actors. Others have been developed by NGOs and do not give adequate attention to the commercial practicality of their application. There may be six, seven or more sets of interests that need to be taken into account in a truly balanced standard-setting process.
- Multiplicity: There is a real problem of loss of effectiveness where there are multiple competing sets of standards. The impact of the system is

---

153 Potts *et al*, "Standards and the Extractive Economy", IISD 2018, www.iisd.org/sites/default/files/publications/igf-ssi-review-extractive-economy.pdf.
RA Hodge, *Towards contribution analysis*, United Nations University (January 2017).

154 www.isealalliance.org/; see also Tracey M Roberts, "Innovations in governance: A Functional Typology of Private Governance Institutions", 22 *Duke Envtl L & Pol'y F* 67 (2011).

often lost if there are too many different sets of standards promulgated by too many different organisations. The nightmare example is the explosion of organisations offering certification of ecotourism operations, which have created a Babel that undermines the effectiveness of everyone.

- Scope of application: Some standards apply to particular products. Others apply to companies. There are some that apply to individual facilities (eg, a company can certify one of its mines, but not others). Some apply to governments. There are some that apply to both companies and government.[155]

(b) *A validation process*

Merely having a standard may provide guidance to parties that are trying to do the right thing, but lack any kind of benchmark against which to judge their performance. And announcing adherence to it may create some reputational benefits. It may be more efficient for companies to adopt existing sets of rules rather than develop their own. And being part of a recognised system may help in recruiting and retaining talented employees.

But having a set of standards seems to have a limited impact if there is no credible way to determine whether someone is or is not complying with the standard. There might be some value in a standard that is applied by companies simply announcing that they adhere to it, without any external validation. But the value is probably minimal. Some companies have internal validation processes. The IRMA system offers a 'self-assessment tool'.[156]

Perhaps the next step is to have an industry organisation evaluate and validate the performance of its members against the standard. One of the most widely recognised approaches in this area is that of the ICMM.[157]

Another approach is to require companies to contract external audits, on the model of a financial auditor. The external auditor of performance would then give a public report of its opinion as to whether the company was complying with the standard.

The most credible and robust approaches seem to be those where the validation process is responsible to an independent multi-stakeholder management. This is an area in which participation of NGOs may be critical: if people from Human Rights Watch or Greenpeace are helping to manage the validation process, it will probably be much more widely accepted than a purely

155 See, for example, the EITI, https://eiti.org/; Aaronson, SA (2011), "Limited partnership: Business, government, civil society, and the public in the Extractive Industries Transparency Initiative (EITI)", *Public Admin Dev*, 31: 50-63. doi:10.1002/pad.588

156 https://responsiblemining.net/

157 See, for example, www.icmm.com/website/publications/pdfs/biodiversity/biodiversity-performance-review_full-report.

company-driven or industry-driven approach. Examples include the validation process of the Forest Stewardship Council[158] or the EITI.[159]

(c) *Incentives and consequences*

Even if we have a good standard and a robust validation process, we are still missing something.

There need to be some rewards for compliance or some sanctions for non-compliance. If the standard is good and a company has passed a rigorous independent verification, there is (or at least should be) reputational benefit.

Reputational benefit alone may entice some potential participants to participate. They may be convinced that there are concrete benefits, such as higher employee morale and easier recruitment, or better relationships in the communities where they operate. Another oft-cited benefit is the reduction of business risk. Compliance may also make their stock more attractive to socially responsible investors.

Compliance with the IFC Performance Standards and the related Equator Principles[160] (subscribed to by most western commercial banks) are supposed to be a precondition to obtaining credit from these organisations. Access to credit, or access on favourable terms, is the incentive.

Preferential insurance rates, granted on the rationale that compliance with standards reduces risk, are also an incentive.

There are major buyers in a number of markets which will only (or preferentially) buy products that have been certified under the appropriate certification scheme. An example: "More than 90% of Walmart U.S., Sam's Club and Asda's (U.K.) fresh and frozen, farmed and wild seafood has earned Marine Stewardship Certification or Best Aquaculture Practices."[161] So the incentive is preferential sale of products to major purchasers.

We should soon see the extent of this effect in the mining industry as IRMA certified minerals become available.

The EU member countries have a system where government purchasing gives preference to certified products.[162]

Some systems target individual consumers. Fair Trade coffee is a well-known example.[163]

---

158 https://ic.fsc.org/en; Wanggi Jaung *et al*, "Forest Stewardship Council certification for forest ecosystem services: An analysis of stakeholder adaptability", *Forest Policy and Economics*, Volume 70, 2016, pp91-98, ISSN 1389-9341, https://doi.org/10.1016/j.forpol.2016.06.004.

159 https://eiti.org/; Benjamin K Sovacool *et al*, "Energy Governance, Transnational Rules, and the Resource Curse: Exploring the Effectiveness of the Extractive Industries Transparency Initiative (EITI)", *World Development*, Volume 83, 2016, pp179-192, ISSN 0305-750X, https://doi.org/10.1016/j.worlddev.2016.01.021.

160 http://equator-principles.com/; Mizuho Financial Group, "What Are the Equator Principles?" www.mizuho-fg.com/csr/business/investment/equator/about/index.html.

161 Walmart: Sustainable Food, https://corporate.walmart.com/global-responsibility/environment-sustainability/sustainable-agriculture.

162 European Commission on Quality Labels, https://ec.europa.eu/agriculture/quality_en.

163 Fairtrade America: Coffee, http://fairtradeamerica.org/Fairtrade-Products/Coffee.

These are not the only ideas – but any serious system needs to have an idea of what incentives it can offer for compliance or what sanctions it can impose for non-compliance.

Again, this element needs to be thought through at least as much as the contents of the standard itself. Without some kind of enforcement mechanism, or incentives, the standard will have much reduced impact.

5. Conclusion

Political pressures, social needs and business imperatives have led to the development of a wide variety of provisions that are intended to bring some order to global trade and investment markets in the minerals industries. The public, investors, mining companies and their critics have all, in different ways, seen this development as meeting their needs and objectives.

The extractive industries have the potential to be a major engine of sustainable development in some of the world's poorest regions, providing needed resources to the world's consumers in the process. Flaws in the legal and policy framework are part of the reason that they have too often fallen short of these objectives.[164]

Some of those flaws can be dealt with in national legislation and policy. But some of them cannot be resolved without a favourable system of international rules and standards. Creating these, in the absence of effective global rule-making organisations, is challenging to say the least.[165] The nature of the global mining industry has placed it in a leadership role in that process.

There is a pressing need to develop the methods to evaluate these various systems, and their governance, to move towards systems that are more effective, more synchronised and at the same time more attuned to democratic norms of governance.[166]

The author would like to thank SDSG intern Conner Sapp, as well as Rachel Amspoker and Miles Williams, for their research support in the preparation of this chapter.

---

164 See L Danielson, "Sustainable Development in the Natural Resource Industries: New Perspectives, New Rules, and New Opportunities", in Rocky Mountain Mineral Law Foundation, Proceedings of the 50th Annual Mineral Law Institute (2004).

165 See Sanjeev Khagram and Saleem Ali, "Transnational transformations: From government-centric interstate regimes to cross-sectoral multi-level networks of global governance" in J. Park *et al* (eds), *The Crisis of Global Environmental Governance: Towards a new political economy of sustainability* (Routledge 2008).

166 See Potts, *et al*, "State of Sustainability Initiatives Review: Standards and the Extractive Economy" (IISD, 2018). See www.iisd.org/sites/default/files/publications/igf-ssi-review-extractive-economy.pdf.

Mining capital finance – traditional and alternative mechanisms

Brian P Dominique
Cassels Brock & Blackwell LLP
Patrick Petzall
Formerly of Cassels Brock & Blackwell LLP

1. Introduction

Mining is a capital-intensive industry, subject to challenges of volatile commodity prices, the fluctuations of local and global commodities and capital markets and economies, significant geo-political risks and enhanced public scrutiny. This environment makes securing the significant amounts of capital needed for mining exploration and development particularly challenging, and has caused the industry to focus in recent years on developing creative financing alternatives.

This chapter provides an overview of how innovations in mining finance developed in Canada, a hub for international mine financing, have enabled the development of alternative forms of mining finance throughout the global industry. It describes the principal financing mechanisms available to junior exploration and major mining companies and discusses the main features of each mechanism.

The chapter also discusses in greater detail the accessibility and structure of the two main alternate mining finance mechanisms – royalties and metal streams – addressing their respective payment structures, covenant packages, security and other risk allocation issues, and outlining some of the most common provisions found in royalty and streaming agreements. It also provides a comparative chart summarising the advantages and disadvantages of each mining finance mechanism.

2. Alternative financing history

Until 2012, mining companies were among the favoured investments of the capital markets and thrived following the financial crisis of 2008-09. By the first quarter of 2013, however, continuing slumping global demand and softening commodities prices caused investor preferences to change and, for the first time in a decade, there were no mining flotations on the Toronto Stock Exchange.[1]

1 Kari MacKay and Mark Bennett, "Under the Rocks are the Words: How a metal purchase agreement revolutionized alternative financing and launched the new majors – a look back at the first decade of metal streaming transactions", *Proceedings of the 60th Annual Rocky Mountain Mineral Law Institute* (2014), p16–6.

During the period from 2007 to 2013, equity financing tightened up (dropping from a high of US$45 billion in 2009 to just above US$5 billion in 2013). Equity financing was so dilutive at the then current share prices that almost no company pursued it. This was in large part due to broad risk aversion in the markets, which brought down commodity prices and turned investors away from volatile mining stocks. It forced mining companies to pursue alternative financing sources, including royalty financing and metals stream financing.[2]

Mining royalties have been around for many decades and have evolved over the years to meet prevailing industry requirements. Early royalties were often created as an ongoing upside for a prospector or mining company wanting to divest of an interest in a mineral property by sale or under a joint venture agreement, but unwilling to give up completely on the property's potential value or unable to accurately value a sale price to reflect mineralisation at the time of divestiture. These royalties emerged as part of the consideration for the acquisition of rights in a mining property. As royalties became assets that holders could sell for valuable consideration, the first royalty company models that purchased/held or sold portfolios of royalties as their main business emerged. By 2005 there were several publicly traded royalty companies worldwide and their business model focused on the acquisition of portfolios of existing royalties.[3] From 2007 to 2013, royalty financing was another element of portfolio optimisation. It is in this period that the beginning of metals stream financing also emerged.

In 2004 Silver Wheaton Corp was formed. It differentiated itself from traditional royalty companies by having a single commodity focus on silver and by structuring its investments in a different manner from royalties, through metals streaming (pre-production purchase of future metals streams), as is described in more detail later in this chapter.

By the time of the equity/debt crisis of 2012–13, the capital-intensive mining industry had to look to alternative sources of capital to make up the difference or to diversify its traditional capital models. Metals streaming continued to grow during this era as an alternative financing mechanism.

3. Mining financing mechanisms

Figure 1 shows the main capital financing options available to mining companies (general descriptions of each main financing option follow).

---

2 "Miners forced to get creative as traditional financing sources dry up", *The Financial Post*, 24 July 2012, http://business.financialpost.com/investing/miners-forced-to-get-creative-as-traditional-financing-sources-dry-up.

3 Nancy Eastman, Brian Graves and Frank Mariage, "Global financing alternatives: a primer on royalty and stream financing", https://gettingthedealthrough.com/area/22/article/29109/mining-2017-global-financing-alternatives-primer-royalty-stream-financing/.

Figure 1. Capital financing options for mining companies

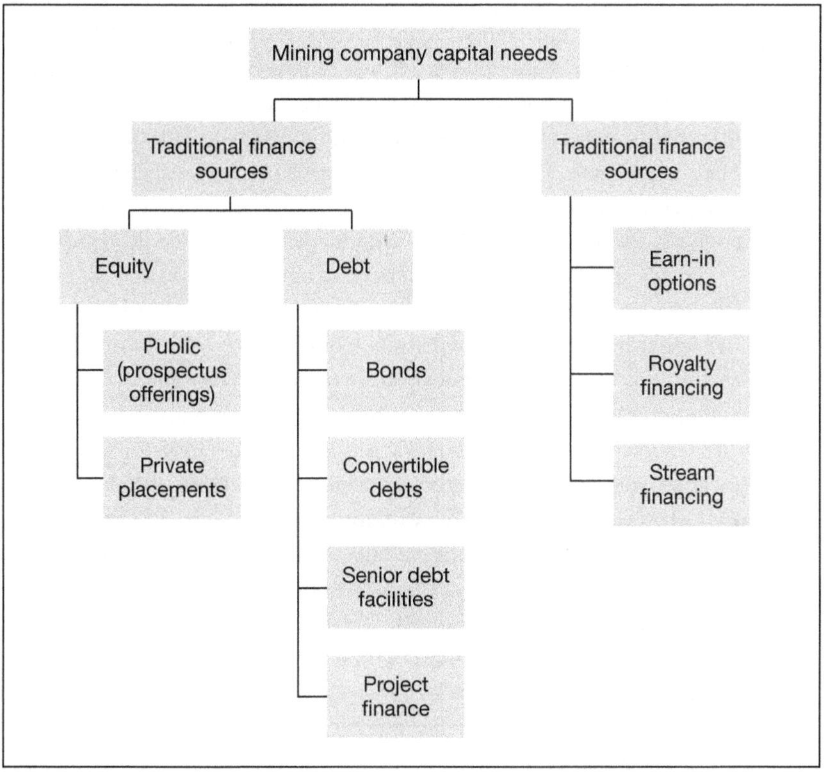

3.1 Equity

Equity is generally the first source of financing in the mining industry, particularly at the exploration stage (limited early cash flow) until the development stage, but can be accessed at all stages (subject to limit on flow-through shares (a Canadian concept whereby rights of deductibility 'flow through' to shareholders) being available only in the exploration stage).

In an equity financing, a company issues shares from treasury for cash (or property) consideration. Shares represent an ownership interest in the company and the shareholder's risk is limited to the value of its investment. Shares may be voting (common) or non-voting, and some may include a right to a dividend (payout of income by the company). Wide share issuances facilitate raising large sums of cash available to the company at lower individual expense and risk to investors, and this facilitates equity financing at early stage mining.

Equity offerings are subject to securities laws and regulation and generally comprise prospectus offerings and private placements.

Prospectus offerings are transactions whereby securities are sold pursuant to a prospectus. Unless a company has an exemption from the requirement to issue its securities under a prospectus under applicable securities laws, the company can only issue securities pursuant to a prospectus offering. A prospectus is a comprehensive registered selling disclosure document prepared and certified by the issuer and the underwriters which describes in detail the company and its holdings (assets, projects), capitalisation and current financial status (including financial statements), future plans and details on allocation of the equity offering proceeds. Prospectus offerings are expensive to the company (incurring underwriters/brokers' fees and supported by legal, accounting and technical advisers) and time consuming to compile, and are subject to regulatory review.

Private placements are transactions whereby securities are sold pursuant to a statutory exemption from the prospectus and registration requirements under the applicable securities regulation requirements – for example, the Ontario Securities Act. Investors under a private placement rely on the company's current and up-to-date filed public disclosure records. The exemption used most often in Canada to enable a private offering is the 'accredited investor' exemption, which includes institutional investors as well as high-net-worth and/or designated sophisticated purchasers. This exemption is found in Section 2.3 of National Instrument 45-106 (and Section 73.3(1) of the Ontario Securities Act).

(a) The special case of Canadian flow-through shares

Canadian mining and resource companies with a Canadian project(s) may issue equity as 'flow-through shares'. Flow-through shares are treasury shares that allow investors to claim certain eligible tax deductions or credits in respect of the mining company's eligible expenditures made on its Canadian mining exploration (and limited development) projects. Flow-through shares are particularly suited for mining and resource companies because they allow certain tax-deductible expenditures to be 'renounced' by the company and to 'flow through' to the purchasing investors.[4] Flow-through shares can be issued through a prospectus offering or a private placement.

The public policy rationale for the introduction of flow-through shares was to use the provisions of the Canadian Income Tax Act to promote early stage mining and hydrocarbons investment in Canada by allowing deductibility against other sources of income for investors; the scheme took advantage of the fact that early stage exploration companies, lacking an income to deduct against, had little need for the exploration and development expense tax deduction.

Department of Finance (Canada) reports have found that flow-through shares have been successful in leading enhanced investment to spending on mining and oil exploration (since extended to clean energy development) and worked to the advantage of junior exploration companies.[5]

4 Norton Rose Fulbright, "A Guide to mining projects and mining finance in Canada" (2013).
5 Canada Department of Finance, "Flow-Through Shares: An Evaluation Report", October 1994, www.fin.gc.ca/pub/pdfs/flowthru_e.pdf.

(b) *Special disclosure for mining projects in Canada – Instrument 43-101*

A key matter that must be taken into account by mining companies seeking equity financing for their projects in Canada is the requirement for securities issuers to comply with the technical disclosure requirements concerning mineral resources set out in National Instrument 43-101 Standards for Disclosure for Mineral Projects.

National Instrument 43-101 was adopted in response to a number of fraudulent schemes in the mining sector involving public companies. Bre-X, which was probably the most famous, involved the large-scale falsification of drilling and assay data. However, there have been others: Naxos (1996), which involved fraud arising out of proprietary assaying; Timbuktu (1996) involving 'salting'; and Cartaway (1996), which involved false 'visual estimates', among others.

National Instrument 43-101 governs the form, content, quality and timing of an issuer's disclosure relating to mining exploration, development and production, creating a uniform and more reliable system of disclosure.

All disclosures of scientific or technical information made by an issuer concerning a mineral project on a material property must be based on information prepared by or under the supervision of a qualified person or approved by a qualified person. A qualified person is an individual, geoscientist or engineer with a minimum of five years' relevant industry experience who is a member in good standing of a recognised professional organisation.

National Instrument 43-101 establishes detailed rules concerning disclosure of mineral resources, setting out the categories of mineral reserves and mineral resources and the different levels of certainty of each (indicated, probable, measured and proven), and sets out detailed rules concerning the preparation and filing of technical reports with respect to a subject mining property, each of which must be prepared by or under the supervision of a qualified person.

Issuers are required to publicly file a technical report in support of scientific or technical information regarding mineral properties that are material to the issuer contained in different types of documents, including prospectus, information circulars, offering memoranda (unless provided only to accredited investors) and valuations required by securities laws, among others.

Technical reports include (ordered by level of specific detail and confidence) preliminary economic assessments, preliminary feasibility studies and feasibility studies. The greater the confidence level, the greater the effort and investment of resources that will be necessary for the preparation of these documents.

3.2 Debt

Mining companies may seek debt funding through the securities markets by issuing corporate bonds or convertible debt. Outside of the markets, mining companies may seek direct financing by banks and other financial institutions by way, for example, of senior debt facilities or project finance.

(a) *Non-convertible debt securities*

A debt security evidences a debt issued by an issuer to a purchasing investor.

The issuer of a debt security borrows money through the markets by selling debt securities, and the investors receive a promise to repay the principal at maturity and interest during the term of the debt security. The holders of debt securities have no ownership rights in the issuer (for a discussion on convertible bonds, see section 3.2(b)).

Bonds and debentures are common features of the debt markets, and their features may vary broadly. These debt instruments may or may not be listed on stock exchanges. Public offerings of equity and debt securities in Canada are subject to substantially the same regulatory requirements and thus an offering of debt securities requires a prospectus, unless an exemption applies. An issuer must file and provide to potential investors both a preliminary and final prospectus and obtain receipts of acceptance for those prospectuses from the applicable Canadian securities regulator.

Issuers of debt securities are generally well-established companies with a good credit rating, as investors are more willing to invest in a company with a good credit history (and strong balance sheet and cash flow) and reputation. If the issuer is not of sufficient size and creditworthiness to be acceptable to the debt securities markets, it will be unable to issue debt securities and will probably have to rely on secured loans from banks or other lenders, or pursue alternative financing options.

Debt securities are most commonly offered by private placement (with an offering memorandum or term sheet). If using the public market, they are most commonly offered via a base shelf prospectus or a short-form prospectus.[6] Debt securities are typically unsecured other than high-yield bonds, which may, in certain circumstances, be secured by second-priority security on the same collateral as the issuer's senior bank lenders typically hold a first-priority security interest.

(b) *Convertible bonds*

A convertible bond is a hybrid security that combines elements of debt and equity securities. The holder of a convertible bond is typically paid a fixed interest rate on the bond until such time as it elects to convert the bond into other securities of the issuer (typically common shares) at a specified conversion rate before maturity.

---

6 Thompson Reuters Canada, *Debt Securities in Canada: Overview*, Practical Law Canada Practice Note Overview.

Convertible bonds in Canada are typically non-investment-grade, are issued without a credit rating, do not include covenant protections similar to other debt securities, are unsecured and are subordinated to senior indebtedness.

The potential benefits for issuers of convertible bonds are as follows:

- The issuer receives upfront payment;
- Investors tend to accept lower rates on the convertible bonds, taking into account the value of the underlying conversion to equity rights; and
- The issuer may be released from its indebtedness if the bonds are converted prior to maturity (eg, if the issuer's stock performance improves).

The challenges of convertible bond issuance are as follows:

- The issuer's ownership is eventually diluted; and
- If the issuer's stock underperforms and investors do not convert before maturity, the issuer will have to pay interest on the bonds during their term and principal at maturity.

(c) *Debt facilities*

A debt facility is a type of loan and may include a revolving credit or term loans, among others.

A borrower may get a loan from a single bank (a bilateral loan) or a group of banks (a syndicated loan). A bank loan is drafted in a form specifically tailored to the borrower's particular needs. The most common types of loans are term loans and revolving credit facilities. Where a term loan typically involves a single borrowing to be repaid in set instalments without reborrowing, a revolving credit facility allows multiple borrowings or draws, subject to repayments over time.

Loans have some advantages over debt securities – mainly that loans tend to have lower interest rates than debt securities and tend to be less expensive transactions to the borrower than issuances of debt securities. However, the availability of debt facilities hinges on the robustness of the borrower's balance sheet and cash flow.

Senior mining companies with strong balance sheets will have funding options that include raising investment grade debt and taking on corporate debt. In contrast, mining companies with speculative or pre-development projects will face significant challenges and will likely be required to assume higher interest rates and more restrictive covenants.[7]

The key challenges associated with credit facilities are as follows:

- the requirement for extensive security packages over the company's assets (which may not be available for junior mining companies and may be risky to a secured lender to a mining company);

---

7 MacKay and Bennett, 16-10.

- extensive books and records access;
- scheduled reports on the company's finances and project progress reports;
- restrictions on other indebtedness and payments (including dividends);
- the obligation to maintain certain financial covenants including debt service, leverage and equity ratios; and
- the rigid cash-flow and repayment structures established to ensure full repayment to the lender. For mining companies with smaller or less experienced management teams, these covenants can be burdensome.

(d) Project finance

Project finance is the financing of long-term projects, based on a non-recourse or limited recourse financial structure, in which project debt and equity used to finance the project are paid back from the cash flow generated by the project. Project finance is structured financing of a specific economic entity – a special purpose vehicle (SPV) – created by the sponsors using equity or debt. The lender considers the cash flow generated from this entity as the major source of loan reimbursement sponsor/promoter. If the borrower defaults, the lender can seize the SPV or the assets of the SPV, but cannot seek out the borrower for any further compensation, even if the SPV does not cover the full value of the amount defaulted.

Cash flows generated by the SPV must be sufficient to cover payments for operating costs and debt service. Only residual funds can be used to pay dividends to sponsors undertaking project finance. Project finance is not easily available to junior mining companies or to cover early exploration efforts. The heavy reliance on a predictable project cash flow clashes with the high resource discovery risk and commodity price volatility typical of early stage mining projects.

Further details of project financing can be found in the chapter on the topic.

4. Alternative finance mechanisms

4.1 Earn-in option (joint venture model)

In an option and joint venture model, a junior exploration company will grant an option to a more established company with available cash to earn a controlling (or lesser or greater (ie, 100%)) ownership interest in its mineral property in exchange for the major mining company funding certain minimum expenditures on exploration programmes, taking on the task of creating and delivering technical reports (as those provided under National Instrument 43-101) or meeting project milestones on the mining property according to an agreed project schedule. Typically, the junior company is not required to fund the initial stages and, as the project advances, there is a point where both parties start contributing to the project in proportion to their participating interest in the project (typically from and after the investor point of earn-in).

Earn-in options are attractive to junior companies because they give them access to funding for their exploration programmes on their properties without having to incur significant debt or dilution of capital at the initial stages. This model is also attractive for the major mining company, as it enables it to source and explore promising early stage mining properties without a large commitment of cash or personnel.[8]

Further details of earn-in options and joint venture agreements can be found in the respective chapters on these topics.

4.2 Royalties

Royalty agreements are contractual agreements, whereby a royalty company makes a one-time upfront payment (as either a financing mechanism or a term of purchase) to a mining company in return for future payments, typically based on a percentage of mineral production proceeds. Royalties involve the right of the royalty holder to receive a cash payment (or metal equivalent) from a mining company or mine operator.

(a) Access to royalty financing

Royalty financing is most appropriate for companies with mining projects with proven reserves that can be reasonably expected to produce predictable returns.

Funding by way of a royalty sale typically will have no impact on the mining company's balance sheet and will result in limited earnings dilution. Also, royalties have no impact on share ownership and would not lead to equity dilution. However, royalties will imply a dilution of the company's asset base as the royalty holder has an actual interest on the mining rights themselves (except when the parties intend to make the royalty just a contractual obligation).

Royalties may be used in structuring purchase, option and joint venture transactions involving mineral properties, or may be granted on mineral properties and sold to generate revenue for the owner/operator, and may be payable on only one mineral, on several minerals or on all minerals.

Royalties are non-participating interests in the mining project. The royalty holder is not expected to make additional investments or payments beyond the upfront royalty purchase price and royalty holders are not commonly granted the right to influence or direct mine development or mining operations.[9]

---

8 Darrell W Podowski, Fred R Pletcher, Kuno Kafka and Keenan Hohol, "Junior Mining Companies in Today's Market, Part 2: The Junior/Major Struggle—Negotiating Mining Option and Joint Venture Agreements", *International Mining and Oil & Gas Law, Development, and Investment*, Paper 9 (Rocky Mt Min L Fdn 2015).

9 Alan L Monk, "Understanding Streaming Agreements and Royalty Agreements: Alternatives to Traditional Financing", *Rocky Mountain Mineral Law Foundation Journal*, No 51, p3.

(b) *Types of royalties*

Production royalty: A production royalty (also known as a 'unit' royalty) is a royalty based on a price per unit of production (ie, ounces, pounds, metric tonnes) without any deductions for costs. The production royalty is typically used for low-unit-value material which can be marketed without requiring extensive processing and which is relatively fungible, including solid bulk minerals such as coal, iron ore (pellet/sinter) potash or other industrial minerals and even waste rock (in demand for backfill and mine rehabilitation).

One benefit of the production royalty is that it is easy to calculate and easy to verify. The production royalty works well if relevant commodities prices are relatively stable.

Production royalties may include price adjustment provisions linked to periodical variations of an index of inflation, or may include a price scale per unit, the price payable from time to time to be determined by the market price of the commodity.

Overriding royalty: An overriding royalty ('gross overriding' or 'gross proceeds' royalty) is usually calculated based on the agreed percentage of the monetary value of metals/mineral products produced, without deduction for costs. The overriding royalty essentially measures revenue with few (if any) deductions.

Both the production royalties and the overriding royalties benefit the payee, in that the royalty is payable regardless of whether production is profitable to the payer.

Net smelter return royalty: Net smelter return (NSR) royalties are really a form of gross royalty and are typically structured as a percentage of revenue received or credited to the mining company by a processor or other purchaser after certain limited and circumscribed deductions, including transportation, sampling and assaying costs.

An NSR royalty usually becomes payable as soon as production begins ('commencement of commercial production' – timing to be negotiated and agreed), but may be subject to a production threshold before it is payable. It is considered as a cost of production in determining the economic viability of a mining project. It may be structured so that it ends after a specified quantity of minerals/metals produced or at a monetary cap, or it may continue for the life of the mine.

One benefit of an NSR royalty is that it is relatively easy to calculate and verify, which makes it less likely to result in the types of calculation disputes more commonly found with a net profit interest (NPI) royalty (discussed below) and, as a result, NSR forms of royalty are more prevalent.

Usually, the deductions are limited to:

- mineral product transport, insurance and security costs for transporting the mineral product from the mine to the intermediary processor and/or ultimate buyer;
- all costs and expenses incurred in connection with the refinement or other beneficiation of the mineral product, including milling (negotiated), smelter and refinery charges and all weighing, sampling, assaying, representation and storage costs, metal losses and any penalties for impurities charged by a processor; and
- any applicable value-added or sales or other similar taxes relating to disposal or sale of mineral product (but not including the payer's income or capital taxes).

Exploration, development and capital and sustaining capital costs and the costs of mining at the site are not deductible, and costs of milling on site (or third-party milling costs) may be a negotiated deduction.

Net profit interest royalty: The NPI is a percentage of the profits realised from operations carried out with respect to the property less operations costs and after all exploration and development (capital) costs have been recovered by the operator. Interest on money borrowed by the payer for the project is also deductible in calculating net profit. An NPI royalty is not regarded as a cost of the project to be taken into account to determine whether a property can be mined economically.

Unlike the NSR royalty, the holder of the NPI royalty will not necessarily accrue payments on the royalty upon commencement of commercial production (however defined), as the payer will be entitled to deduct all exploration and capital costs (which may be large) prior to earning a 'profit'.

NPI royalties are typically difficult to calculate and may attract significant disputes related to timing of commencement of payments and proper calculations of payments. Meeting of the minds and precise language in respect of entitlement of payer to make deductions, timing of deductions and applicable accounting procedures and standards are critical.

(c) ***Structure of royalties***

Royalty agreements are generally for the life of a project, although they may include a payer buy-out clause or may include a monetary or production level threshold and/or cap. Uncapped royalty agreements can offer significant upside to an investor if a project proves to have more or other mineralisation than initially contemplated.

Covenants: Given that the upfront consideration made by the royalty holder to the payer is not required to be repaid, royalty agreements do not include extensive covenant packages similar to those found in debt structures. Royalties usually do not include penalties for construction or completion test delays or for failure to meet operational thresholds for funding or solvency ratio tests, and royalty agreements tend to have lesser financial, legal and informational covenants than those found in debt or (in some cases) in streaming structures. However, royalty holders will require regular reporting by the payer and access to books and records of the payer to ensure that they are receiving their royalty share of mine production and to confirm royalty calculations (including interest).

In addition, a royalty holder that contemplates selling its interest in the future may wish to seek a right to obtain broader access to information about the project, not just the financial information necessary to verify the royalty payments, as this information will be useful to justify the value of a royalty to a potential purchaser (particularly a royalty company).

Common royalty provisions: The royalty agreement can provide the royalty holder with the right to obtain payment in the form of a portion of the minerals extracted from the property known as 'taking in kind'. Usually, such rights are limited to a one-time option or limited number of rights to exercise with substantial written notice to the payer, to allow time for direction and redirection of metal flows.

Parties may agree to have advance royalties to be paid periodically until commercial production commences, with advances made usually credited against actual production related royalty payments later due and payable. These provisions are aimed at promoting property development and discouraging the mining property holder from engaging in property warehousing.

Mining companies may reserve the right to buy back all or part of the royalty; this right can be tied to a specific overall amount and is usually limited in time – for example, providing that the buy-back right may be exercised only before commencement of commercial production.

Royalty agreements usually contain provisions that acknowledge and preserve the discretion of the operator over exploration, development and mining decisions.

Security: Security of a royalty holder's interest on title to the property in the form of a collateral mortgage/charge or on proceeds under personal property security legislation is a matter of negotiation and is important to protect the right of the payee if there is a sale of the property or if the royalty is in arrears at the time of a payer insolvency.

Right of first refusal: Royalty agreements may contain a right of first refusal granted by the payee to the payer, which gives the payer the right to acquire all or part of the royalty interest for the amount offered to the payee by a third party. This is potentially an important 'buy-back' right of the payer, but may devalue the market value of the royalty for the payee. Such provisions usually require a payee that has received an offer from a third party (which the payee is willing to accept) to offer such royalty interest to the payer on the same or substantially similar terms. The payer will have a fixed period to accept the offer and if it declines the offer, the payee is free to sell that interest in the royalty to the third party on the same terms within a specified period of time.

Royalties as an interest in land: A key feature of royalties is that the common practice is to make royalty interests follow the property. In this regard, in Canada, the Supreme Court of Canada in its 2002 decision in *Bank of Montreal v Dynex Petroleum Ltd* established that a common law royalty interest can be an interest in land as long as the following test is met:

1) *the language used in describing the interest is sufficiently precise to show that the parties intended the royalty to be a grant of an interest in land, rather than a contractual right to a portion of the substances recovered from the land; and*
2) *the interest, out of which the royalty is carved, is itself an interest in land.*[10]

These criteria have been applied most recently by the Ontario Court of Appeal in its 15 March 2018 decision in *Third Eye Capital Corporation v Ressources Dianor Inc.*[11]

Two points are key:

- Not all royalties are inherently interests in land under common law; to secure that character, the parties must make their intention clear in the relevant royalty agreement; and
- The principles of this rule apply only to royalties under common law; they will not necessarily apply to royalties subject to civil law, as that real property regime is separate and distinct from those of the common law jurisdictions.

4.3 Metals stream financing

Metals streaming transactions are purchase and sale agreements that provide for the purchase and sale of the applicable commodity – specifically, of the specified metals or mining products produced. In a streaming transaction, the mining company agrees to sell, and the streaming company (the stream purchaser) agrees to purchase, a certain percentage (or all) of one or more metals/minerals produced from a mining operation at a fixed price, which price is typically below the prevailing market price and usually approximates the cost of producing and delivering the minerals purchased.

Streaming transactions are characterised by the physical purchase of future

10 *Bank of Montreal v Dynex Petroleum Ltd* 2002 SCC 7, 2002 CSC 7
11 2018 CarswellOnt 3694, 2018 ONCA 253, 290 ACWS (3d) 20, 57 CBR (6th) 171.

mine production (as opposed to eventual revenues) and are not commonly structured to bring about an interest in land (although they are often secured with a collateral charge on the property). These features, along with the recurring payments for future deliveries (once the deposit has been compensated) distinguish stream financing from royalty arrangements.

The main advantage of a streaming transaction to a mining company is that the streaming agreement enables it to monetise the streamed metals/minerals prior to extracting them. The mining company receives a large upfront payment (or payments) that may be used to build or expand its mine without having to make periodic payments of principal and interest as with a debt transaction, or without its existing shareholders suffering the dilution of their equity participation as may occur with equity financings. A streaming transaction may also be seen by the market as an endorsement by way of investment by the stream purchaser in the mining company or the mine.

The streaming company agrees to make an upfront payment in the form of a financial deposit (or series of financial deposits) in exchange for the delivery by the mining company of an agreed percentage of metal/mineral equivalent actually mined from the relevant mining property for the payment of a pre-established purchase price per unit of metal/mineral delivered.

By structuring the upfront payment as a deposit, streaming companies ensure a return of that portion of their investment either during the term of the agreement or upon termination. The duration of streaming transactions is usually very long term (decades) or life of mine. Typically, the metal/mineral sold under a streaming agreement is a by-product of the mining operation.

Streaming agreements usually provide that if the fixed price paid by the stream purchaser is less than the market price of the metal/mineral, the difference between the market price and the fixed price is 'credited' towards the deposit; once the deposit has been credited down to zero, the stream purchaser will continue to purchase the streamed metals/minerals at the lesser of the fixed price and the market price for the remaining term.

Payment of the upfront deposit can be structured so as to be staged during the development of the project, taking into account relevant financing or construction milestones, and may provide for repayment of the deposit in case of failure to meet certain development milestones or if the project does not reach completion.

Streaming agreements usually provide certain conditions precedent to payment of the deposit, or of any particular portion thereof if payment of the deposit is staggered over time, as a means to ensure proper use of funds for project development; conditions precedent include delivery of security agreements, delivery of inter-creditor agreements (if required) and, if applicable, delivery of a commingling plan.

The stream purchaser may be given the right to terminate the streaming agreement if the conditions precedent to payment of all or any part of the deposit are not met. A streaming agreement will usually require repayment of the deposit upon early termination of the agreement under certain circumstances, such as breach, and for this reason the deposit should be secured with a collateral charge against the property.

A streaming agreement may require delivery of actual metal/minerals or, more likely, a credit to the streaming company's designated metals account. When the streaming agreement provides for the mining company to credit the streaming company's metal account, no actual product must be delivered, but the mining company will need to transfer metal market credits to the streaming company's account in lieu of physical delivery of actual product.[12]

Streaming agreements may also include provisions prohibiting employee solicitation by the stream purchaser and standstill provisions prohibiting the stream purchaser from acquiring a certain threshold percentage of shares of the mining company. The mining company is also usually required to provide copies of all offtake agreements to the streaming company together with a covenant to enforce them.

(a) Mineral products

Initially, streaming models were applied to precious metals that were the by-product of the main extraction process, as this offered an opportunity to the mining company to realise value (leverage) from by-products that were not its primary source of value. In addition, a base metal mining company generally might not receive the same valuation for precious metal by-product revenue that a precious metal mining or streaming company would. Mining companies are less attracted to streaming their primary metal products and will focus on metal sales or offtake agreements for those primary commodities.[13]

(b) Covenants

Streaming agreements on mines in construction typically include conditions that need to be satisfied before the upfront deposit is paid, such as completion of required permitting and securing construction financing, and also generally include operating covenants to ensure that:

- offtake terms and conditions are commercially reasonable (because they typically trigger delivery obligations);
- the processing (or commingling) of ore from mines outside the scope of the project is limited; and
- the mining company mine plan is implemented so that it extracts resources as if it had a full ownership interest in all resources being extracted.

---

12 Monk, p9.
13 "Under the Rocks are the Words" (*ibid*) pp16–12 to 16–14.

(c) ***Use of funds***

Generally, the terms of the streaming agreement establish the covenants of the mining company with respect to the utilisation of the deposit delivered by the streaming company. In some cases the operator may spend the deposit as it sees fit; but in cases where the purpose of the stream purchase is to finance development, the relevant agreement will provide that the deposit must be spent only for specific purposes – usually towards the construction and development of the project. Such payments are often staged over the development of the project, based on particular construction and financing milestones.

(d) ***Right of first refusal and right of first offer***

Streaming agreements may contain a right of first refusal granted by the mining company to the stream purchaser, which gives the stream purchaser the right to acquire any royalty or mineral stream, or a participating interest in the minerals subject to the streaming agreement that are produced from the property, for a price offered by a third party to the mining company which the mining company is prepared to accept.

The right of first offer is granted by the mining company to the stream purchaser by requiring the mining company to offer the royalty, mineral stream or participating interest that it may wish to sell to the stream purchaser for a certain price first; only if the stream purchaser declines the offer may the mining company sell the relevant interest to a third party for that same price.

(e) ***Access to stream financing***

Generally, streaming companies invest in projects with established mineral reserves/resources, in development projects that need additional financing for construction or in producing mines seeking to expand production. However, streaming companies have been willing to invest a portion of the initial payment to facilitate, for example, feasibility studies.

(f) ***Notes on disclosures***

Securities filings: If a publicly listed company is involved in royalty or streaming transaction, it may be required to disclose the existence of the royalty or streaming agreements, and potentially file the agreements themselves on certain publicly accessible websites such as SEDAR (Canada) or EDGAR (USA). In Canada, an agreement material to the company that is not entered into in the ordinary course of business will generally need to be filed.

Usually, a financing royalty agreement or streaming agreement is not entered into in the ordinary course of business for a mining company; however, such agreements could fit into this category (and thus be exempt from the filing and disclosure requirement) if the listing company is instead the royalty holder or the stream purchaser.

National Instrument 43-101: Pursuant to National Instrument 43-101, an issuer in Canada is required, under certain circumstances, to prepare and file on the SEDAR.com website a technical report on a mineral project that is material to such issuer. A 'mineral project' is defined in National Instrument 43-101 to include "a royalty or similar interest", which is considered to include an interest in a commodity stream from a mining operation. National Instrument 43-101 provides certain limited exemptions from the requirement for a royalty holder or stream purchaser to file a technical report, because it is recognised that it may not have an unfettered contractual right to access the scientific and technical data on the property that is necessary to prepare a technical report.

Some royalty and streaming agreements may specifically include provisions that require the mining company to assist the royalty holder or the stream purchaser to obtain scientific and technical information sufficient for the production and filing of a technical report required by National Instrument 43-101.

5. Advantages and disadvantages of alternative financing mechanisms

Figure 2. Advantages and disadvantages of alternative financing mechanisms

Finance mechanism	Advantages	Disadvantages
Equity	No repayments. Flow-through shares (Canada) incentivise exploration and development funding.	Ownership dilution. Highly regulated. Complex and costly disclosure requirements. Availability may be limited by fluctuations in capital markets.
Bonds	Generally unsecured. No asset or ownership dilution.	Disclosure requirements similar to equity financing.

continued on following page

Finance mechanism	Advantages	Disadvantages
Debt facility/project finance	Lower interest cost to well-established mining companies as compared to juniors. No ownership dilution. Project finance limits mining company's exposure.	Requires a broad and restrictive security package. Restrictive covenant packages. Higher transaction costs. Availability limited by strength of balance sheet and cash flow. Lenders are risk averse. Limited availability for early-stage (exploration) projects.
Earn-in option	Available for early exploration stages. Junior company benefits from leverage and expertise of major.	Ownership dilution. Joint venture partner deeply involved in (control) development and operations (loss of independence).
Royalty	Upfront financial infusion. No pre-established volume deliveries. Little impact on balance sheet. More limited covenants on use of funds. Royalty holder has limited input on operations. Can grant royalty holder an interest in mining property.	Asset dilution. Restrictions on dispositions. Limited availability for exploration stage. No subsequent funding from royalty holder. Terms of royalty holder interest in mining property can be restrictive on charge or transfer.

continued on following page

Finance mechanism	Advantages	Disadvantages
Streaming	No deliverables if production not achieved. Low transaction costs and rapid implementation (two to four months). No ownership dilution. Company retention of control of project/interest alignment between streaming company and mining company.	Dilutive at asset level. Variable financial benefit during life of mine depending on difference between actual and expected production. May affect financials and effect negative cash costs by reducing by-product credits against costs for primary metals. Transfer restrictions.

6. The future of mining finance

Market analysis confirms that the weakening of metal prices from 2011 to 2015 narrowed the availability of traditional forms of mining finance (equity, debt and project finance) to those established major mining companies with well-developed projects and predictable cash flows.

Alternative financing mechanisms based on earn-in agreements, royalty financing and metal streaming became more popular during this period for smaller players for which traditional debt or equity financings have become more difficult.

Analysts expect that, with some recovery of minerals prices, equity and debt markets will become less constrained; but they also expect that the number and size of financing deals will be reduced as, with greater liquidity, major mining companies will likely look to fund their own projects (including through joint ventures with other major mining companies) and projects at the exploration stage, while well-established and proven companies with mining projects at the development stage (where feasibility study financing is needed) are expected to continue to turn to funding through private equity, earn-in options (particularly for exploration) and alternative funding mechanisms, including royalties and streaming.[14]

14 Richard Jansen Van Vuuren, "Mining Finance: equity and debt financing remain key; alternative funding to grow", *Mining Review*, www.miningreview.com/mining-finance-equity-debt-financing-key-alternative-funding-grow/.

Project finance in the mining industry

David Budd
Carla Potter
Cassels Brock & Blackwell LLP

1. Overview

One of the most important factors in the initial success of any mining project is how the construction of the project will be financed. Project finance transactions – being transactions where funds are raised to finance the construction and ultimate build of the mine to achieve operational status – are multi-faceted processes which differ from traditional bank financing due to the inherent uncertainty as to whether the project can be built on time and on budget. In structuring and organising a mine project financing, advisers must consider the unique factual circumstances of each project, which encompass everything from the risks associated with the jurisdiction in which the mine operates and the experience of the executives leading the mine project financing to the regulations affecting the project and the stakeholder groups that may have a vested interest in, or oppose, construction of the mine. Such factors make each mining project, and thus each project financing, unique. In addition, advisers must be attuned to the many possible issues that can arise in such a financing to avoid undue delays that could jeopardise the build of the project or make the project no longer economically viable. These issues also differ depending on the unique factual matrix of the project and the jurisdiction in which the project is located.

As an overview, project financing in the mining sector generally involves a combination of debt and equity which is raised in order to build a project to completion and ultimate revenue generation. Project financing does not finance exploration and development activities undertaken in connection with determining whether a project is viable, but rather focuses on the construction of a mine that is already determined to be economically feasible. Depending on the economics of the project and the resilience of the financial markets in the mining sector, the amount of capital and the sources of financing may vary and may include, to supplement the debt and equity financing package, alternative financing arrangements such as royalty agreements, stream agreements and offtake agreements to get the construction of a project fully financed. Debt financing is generally sourced through both traditional banks which operate in the mining space and alternative funds which may provide more operational flexibility for the project entity in return for a higher cost of capital.

Project financings in the mining sector are generally done on a secured basis, whereby on the occurrence of a default under the debt facility, the lenders have both asset-level security to realise on the specific assets of the project entity – which is itself generally a special purpose vehicle (SPV) – and a pledge of the shares or other equity interests in the project entity in order to allow for control of the project during the realisation process. By taking security, lenders have more avenues of recourse ahead of other creditors in a realisation scenario. In addition to security from the project entity, guarantees and security are generally provided from a project sponsor which is often the publicly listed parent of the project entity, as well as often other entities within the project entity's corporate group.

This chapter explores these key aspects of project financing in more detail, and reviews at a high level unique aspects of project financing transactions in the mining industry and some possible pitfalls that may be encountered.

2. Project entity

The hallmark of a mining project finance transaction is that the project assets to be financed are bifurcated from the broader corporate group and held within a project entity. This project vehicle (an SPV) is generally a single-purpose entity, usually a corporation or limited partnership, created solely for the purpose of carrying out the construction and operation of the mining project. Controlled by project sponsors, the SPV is the centre of the mining project through its contractual arrangements with operators, contractors, suppliers and customers. The SPV will be dependent on revenue streams from the project, realised through the sale of the saleable product (eg, gold, silver), which will commence only once construction of the mining project has been completed and the project is in operation. As such, there is inherent risk for any lender in a mining project transaction, which continues through to project completion (and often beyond as production ramps up), as the SPV is not self-sustaining from a revenue-generation capacity until project completion is achieved and the product can be mined and sold. In a project finance transaction, the SPV will generally be the borrower under the debt finance component of the project financing, providing the representations, warranties and covenants, and will be bound to the terms of the loan documentation and related security agreements.

The main advantage to a project entity of using an SPV in a mining project finance transaction is that it allows for the separation of the assets and liabilities associated with the project from the assets and liabilities on the corporate balance sheets of the project sponsors, which may be larger companies with potentially more than one mine or exploration property. Without such separation of assets, a default on the project loan may enable the lender to seek remedy against all assets of the project sponsor. By using an SPV, however, the recourse on a loan default is limited to the assets of the SPV, allowing a corporate group with multiple mining projects or exploration properties to carve out non-project related assets from the project lenders' security. While the benefit of such bifurcation is obvious to a large corporate group, practically speaking there may not be as obvious a benefit in using an SPV for a corporate group that is a single project entity before project completion. Furthermore, although the utilisation of an SPV can significantly limit the liability of project sponsors in large corporate groups, lenders often require numerous other guarantees and assurances from sponsors to supplement this limited liability, which in turn negates some of the benefits. These additional guarantees are examined later in this chapter.

3. The sponsor and its role

A key feature of a mining project financing transaction, which differs from a traditional secured bank financing, is the role of the sponsor. Project sponsors are generally the indirect owners of the mining project and those with an equity stake in the project who are viewed by the lenders as the key credit support. The project sponsor is often perceived as the creditworthy entity (as compared to the SPV), which is listed on a stock exchange and thus has access to equity raises in the capital markets to support the debt assumed by the SPV. In addition to generally being required to provide equity to the SPV and thus the mining project to support the financing of the mining project, sponsors may play a role in providing various guarantees, comfort letters and other types of support throughout the life of the project, which will give the lenders comfort as to their ability to recover their investment in an insolvency scenario.

3.1 Equity contribution agreements

Part of the consideration for any project financing will be how much equity is required such that the leverage at the project entity level is acceptable to the lenders. The amount of equity that lenders will require as part of the financing of the mining project is generally dependent on both the economics of the project and the general state of the equity markets (and in turn, the ability to raise equity). To gain comfort on the certainty and amount of the equity injection, project lenders will look for equity contribution agreements whereby shareholders or project sponsors agree to inject equity into a project company by way of either capital contributions or deeply subordinated debt that is payable only following repayment of the project debt. The agreement may specify certain points in time at which the equity must be injected or may specify certain circumstances that would trigger an equity injection – such as upon the occurrence of a default under the project loan facility or to cover cost overrun shortfalls. The specific distribution and frequency of these equity injections typically depend on the creditworthiness of the shareholders or sponsors and the overall capital needs of the project. These contracts are generally entered into by the shareholders or sponsors directly with the lenders or the lenders' agent, and should contain a direct covenant to pay in favour of the lenders or their agent on the occurrence of a default

3.2 Completion guarantees

A completion guarantee in a mining project finance transaction is a generally a commitment by the sponsor or other members in the project entity's corporate group, and less frequently a third party, to complete or provide the funds necessary to complete the mining project through to project completion (ie, when the party begins to produce saleable product and has the ability to generate revenue). While it is often negotiated that the project guarantee of the sponsor falls away on achieving project completion, the lenders may require that the sponsor guarantee extend beyond project completion, and it often becomes a limited recourse guarantee with recourse limited to the shares of the project entity owned by the sponsor such that the lenders retain their option to control the project in an insolvency scenario. These guarantees provide added assurance and a financial commitment that the mining project will be completed on the agreed terms, barring any extreme adverse circumstances. It is unsurprising, then, that an exact definition of 'completion' is a key aspect of both the completion guarantee and the mining project finance transaction, which must be agreed by both parties. While the parties are free to negotiate the exact terms of completion, the typical criteria for completion include:

- confirmation that the project is complete enough to be put to use based on agreed operational thresholds for production;
- demonstration that the project meets the proper performance criteria; and
- demonstration that the project can operate at agreed levels over the term of the loan.

3.3 Cost overrun guarantee

A cost overrun guarantee is a commitment by a sponsor, or less often a third party, to cover any shortfalls that arise if the costs of completing a mining project exceed the project budget agreed by the project lenders. Typically, cost overruns arise in one of two ways. First, cost overruns may arise in the construction phase of a mining project if contracting is not fully complete at the time of financial close or as a consequence of contracts not providing for price-certain terms. These risks are mitigated through the use of a reputable engineering, procurement and construction (EPC) contractor and a negotiated EPC agreement for the mining project, which clearly defines the terms on which the project will be constructed. Second, cost overruns may arise as a result of change orders, which arise if any modifications are requested to the original construction plans for the project or if other unforeseen events occur on the project. Sponsors will often try to limit their obligations to cover cost overruns to an agreed maximum, as this is a more limited degree of liability than entering into a completion guarantee, which is generally an unlimited guarantee before project completion. A cap on any cost overrun guarantee will have to be negotiated with the lender and the ability to negotiate a cap will depend on the other types of support that the sponsor can offer, as well as its creditworthiness.

3.4 Other forms of sponsor support

In addition to guarantees and equity contributions, lenders look to the sponsor as a key piece of the lenders' security panel, to ensure that the lenders can take control of the project in an enforcement scenario. As such, it is typical in a mining project financing transaction for a sponsor to provide a pledge of the shares of the SPV owned by the sponsor, as well as often general security over its assets as recourse for the project lenders. As previously noted, once the project reaches completion, the security that the sponsor provides may either fall away or be limited to the pledge of the shares of the project entity owned by the sponsor.

4. Timeline for project financing transactions

Project financing transactions generally take longer than traditional bank financings of an established business – in part due to the required technical due diligence and risk assessment of the lenders, given that no tangible assets are readily available to realise on in an enforcement scenario until the project is complete, by which time the lenders have extended the full amount of the loan. Furthermore, project entities or their sponsors which have historically only raised equity should be aware that due to the continuing nature of the obligations and the relationship between the lenders and the project entity, the project financing transaction will be significantly more involved and will thus require more time than an equity raise on the public markets would generally take. The timeline for any project financing transaction is very fact specific – for example, a project with a significant requirement for equity may take longer due to market pressures which delay the equity raise, while a project that requires consent from a governmental body to proceed may take longer due to the bureaucracy of obtaining that consent. It is thus important at the outset of a project financing transaction to identify long lead time items that could potentially cause undue delay and thus cost overruns for the project or negative news on the public markets, as a result of the delay in the financing. Seasoned project finance counsel can assist in mitigating these delays and the costs associated therewith.

5. Funding sources

5.1 Overview

Mine project financings are typically conducted through a combination of debt and equity. As previously noted, dependent on the appetite of the markets and the economics of the project, an alternative finance source may be required to ensure the project is adequately capitalised, such as a stream, royalty or off-take agreement. Borrowers will look for the lowest cost of capital with the greatest operational flexibility; but how the project is ultimately financed will often be in part determined by factors outside the control of the project entity and will depend on the appetite of the market at the time of the financing.

5.2 Typical financing sources

The typical sources of financing in a mining project finance transaction are outlined below. How the project will be financed and the overall cost of capital will largely depend on both timing and project-specific factors:

- Equity: Generally, mining project finance transactions require an equity injection, whether from the public shareholders of the project entity or the sponsor thereof. The quantum and timing of the equity injection are generally fact specific and are negotiated between the project lenders and the project entity.
- Bank debt: Traditional bank lenders which are familiar with the mining space are frequent providers of indebtedness in mining project financing transactions. Banks can often provide the most attractive pricing on borrowed money, but may have increased administrative requirements for the project entity, such as increased reporting and compliance requirements during the term of the loan. The requirements of a bank will depend on the conditions that its internal credit department put on the loan, and as such any financing will be subject to compliance with the credit committee requirements.
- Development banks: Depending on the jurisdiction of the project, viable lenders may also include development banks with a mandate to foster development in certain geographical areas. Examples include the European Bank for Reconstruction and Development, the Caribbean Development Bank and the International Finance Corporation. Given that these banks have specific mandates, the access to capital from these banks and the viability of the lending relationship will depend on the development impact of the mining project and the jurisdiction in which the mining project is located.
- Non-traditional lenders: Certain funds have been established that are viewed as alternative lenders in the mining project finance space, as they are not traditional bank lenders. These funds often provide debt

financing at a higher cost of capital than traditional bank lenders, but may offer more operational flexibility to the project entity, which may be attractive to the project entity as it allows companies with limited resources to focus on building the project rather than administrative requirements of the loan documentation. Some alternative lenders bundle their debt financing with other requirements, such as warrants, stream agreements, call options or off-take agreements, as discussed later in this chapter.

5.3 Other key financing sources

Other key financing sources include the following:

- Cost overrun facility: There is an inherent risk in project finance transactions that costs will deviate above the budget and generate cost overruns, which creates a risk as to the project getting to completion and the project lender being repaid. A cost overrun facility is a loan facility available to the project entity to fund such overruns. This facility may be part of the project finance facility with the project lenders, with different terms or pricing, or may be a facility with a separate lender. To the extent that it is a separate facility, the terms will have to be acceptable to the project lender, with inter-creditor agreements in place outlining when the cost overrun lender can be paid and providing that the project lender will control an enforcement scenario.
- Equipment finance facility: Generally, as part of the build of any mine, a significant amount of heavy equipment, from boom trucks to excavators, is required to complete the build. Project entities often address this key financing piece through equipment finance facilities with reputable equipment lessors in the mining industry, such as Caterpillar Equipment Finance. Lenders often look to see that the financing for the equipment finance facility is in place concurrent with the finalisation of the project finance facility, and inter-creditor agreements will generally be required between the project lenders and the equipment financiers.
- Stream agreements: Stream financing is often used by mining projects when raising sufficient debt or equity proves challenging, and as such may be a financing source which is used to supplement debt and equity financing in a project finance transaction. At a high level, a stream financing is a metals purchase arrangement whereby the streamer (the source of the funds) provides a large upfront deposit to the project entity for the right to buy a percentage of future production at a later point in time at a predetermined price, which is often negotiated to be significantly lower than prevailing market rates. This upfront deposit is often attractive for junior mining companies which may struggle to raise

adequate capital, but is less attractive to other potential financiers of the mining project, as stream agreements are long-term agreements with delivery requirements and less than market prices often spanning the full life of the mine.

- Royalty agreements: Royalties are ongoing economic interests in the production or future production from a mining property, with the royalty agreement governing the exact terms of such interest. A royalty typically involves the right of the royalty holder to receive a cash payment from an owner of some portion of the proceeds from the sale of the minerals to which the royalty relates. Increasingly, and more commonly with non-traditional lenders, royalty agreements and other financing sources are paired with the debt financing as part of the capital structure of a mining project financing transaction, such that the project lender is entitled to an ongoing royalty from the mining project long after the project debt has been repaid.
- Off-take agreements: Off-take agreements are agreements between the producer of a resource and a buyer of the resource to sell/buy all of, or portions of, the producer's future production from a specified facility. These agreements are often used in the mining industry where capital costs to extract resources are significant and a company wants or needs to ensure that some or all of its products will be sold at agreed prices. Off-take agreements are often negotiated prior to the construction of a facility in order to secure a market for the future output of the facility; if potential lenders or equity investors can be sure that the company will have a purchaser of all or a significant part of its production on market terms, this makes it easier to obtain financing to construct the facility. Project lenders may look for off-take agreements to be in place in order to have certainty of the sale of the output of the project – a key component in the overall return of the project and repayment of the lenders.

6. Role of advisers

Project financing also differs from other forms of financing is terms of the roles of various third parties, known as advisers.

6.1 Financial adviser

Financial advisers play an important role in providing integrated commercial and financial advice across all stages of a project finance transaction. This is particularly key for junior project companies which are completing their first project financing and generally have both less experience in these transactions and limited administrative resources. At the initial phase of a project finance transaction, the role of a financial adviser typically involves:

- advising on the amount of equity that will need to be raised based on the specifications of the project;
- assisting with the completion of any feasibility studies and financial models associated with project; and
- working to determine the optimal structure and financing method for the project.

As the transaction moves along and lenders have committed to financing a project, the role of the financial adviser shifts to reviewing term sheets and performing financial due diligence, conducting market analysis based on comparator projects and working through the credit approval process with the lenders. Finally, when a project enters the documentation stage, the financial adviser is responsible for assisting in the negotiation of loan documents alongside the legal team and helping to satisfy the conditions precedent for closing and drawing under the project finance facility. Once the project finance facility is in place and all conditions precedent are met, the financial adviser has generally fulfilled its role in the transaction.

6.2 Independent engineer

The role of the independent engineer is to act as a facilitator, and a truly independent participant, which bridges the gap between the financial and operating entities. From the perspective of financial institutions, the role of the independent engineer is to assist in developing reasonable project completion covenants that protect lender interests, and to provide an objective report on the project plans and models relating to construction provided by the project entity. From the perspective of the operator of the project, the independent engineer must help to ensure that covenants allow for reasonable construction timelines and reasonable physical and production completion requirements.

Most of the work completed by the independent engineer focuses on conducting due diligence on bank-level feasibility studies that have been prepared by the mining company or its agents. These due diligence reviews generally cover areas such as:

- the geology and resource estimates of the project;
- the mining and reserve estimates of the project;
- the processing information;
- the environmental and social evaluation, and the permitting for the project;
- the project infrastructure and administration;
- the project schedule; and
- the project economics.

Lending institutions typically request assistance in defining project completion covenants and construction contracts, and the independent engineer is typically retained to monitor construction progress to ensure that construction keeps pace with the capital outlays outlined in the relevant agreements and the operative project plans and models on which the lending institution has relied. The independent engineer also plays a role in determining what constitutes the physical/mechanical completion of the project and the performance/production completion of the project, which is key for achieving project completion under the loan documentation, whereby covenants may relax for the project entity and repayments may begin for the project lender.

It is not uncommon for the independent engineer to continue monitoring the project until the loan has been fully repaid, which may involve providing monthly reports and conducting periodic site visits as required by the lenders.

6.3 Insurance adviser

In most cases, the financial institutions lending in a mining project finance transaction will request the services of an insurance adviser. The insurance adviser helps the parties to navigate a number of different issues, most of which appear greater from the perspective of a prospective lender relating to the insurance programme of the project entity. These issues include the following:[1]

- the appropriateness of the overall insurance package, including the endorsements and quantum of coverage;
- the pricing of the insurance package at different phases throughout the project;
- the possibility for the insurers to cancel policies should certain conditions arise;
- the possibility that sponsors may not renew or may reduce the insurance programme;
- the possibility that the insurers may claim that they were not correctly informed of project risks when these were underwritten, and will therefore reduce or cancel the extent of contractual guarantees;
- the possibility that the sponsors may not pay for insurance coverage or may fail to utilise any claim reimbursements to reinvest in the project; and
- the possibility that the claim reimbursements may be paid to parties that are not entitled to them and that may not make appropriate use of them.

A further aspect to consider is that it is not always possible for a single insurer to cover the entire insurance package, particularly where unique

1 Sefano Gatti, *Project Finance in Theory and Practice: Designing, Structuring, and Financing Private and Public Projects*, Academic Press 2013, pp106–108.

endorsements are required or the project is located in a foreign jurisdiction. This means that there may be different insurance platforms for various insurable risks or entities, and these policies must be understood and reconciled to ensure there are no gaps in coverage. Whenever possible, the use of a lead insurer is generally preferable for sponsors and lenders, because its status is very similar to that of the lead arranger in syndicated loans. The presence of a lead insurer creates a single reference point for full insurance coverage and the ability to have control over the full insurance contract.

In order to address these potential issues, the insurance adviser will review all relevant sections of project agreements and related contracts, including the loan agreement and material contracts that are relevant to insurance, and will engage in discussions with both the lenders and other project parties to gain a complete understanding of the risks associated with the project. The insurance adviser will then confirm whether the insurance coverage that is in place for the project is adequate and sustainable, and whether significant risks remain that fall outside coverage and for which separate endorsements may be required. The role of the insurance adviser often extends beyond the project financing and throughout the life of the project. While the project is operating, the insurance adviser often plays a monitoring role and reports on any cancellations or reductions in an insurance policy, any significant claims and any proposed changes in coverage.

6.4 Model auditor

The model auditor plays an important role in conducting due diligence on the financial models used in project finance transactions, with model audits mainly developed to eliminate errors in the financial modelling that underlies the economics of the project finance transaction. Generally, the financial model outlines the economic viability of the project, showing how cash flows from the project will ultimately result in the project debt being repaid. As such, the financial model is subject to a high degree of scrutiny from the lenders and will generally be subject to covenants in the project loan agreement, including that the financial model cannot be amended without the consent of the lenders. Typically, model audits are requested by banking organisations in order to reassure lenders and investors that the financial models used in the transaction – and which are generally prepared by the project entity, with the assistance of its financial adviser – are accurate and dependable. Model audits review:

- the model's logic;
- the consistency of the model against contractual and financial documents;
- the consistency of the model against generally accepted accounting principles applicable to the project company; and
- the sensitivity of the model.

These audits may take anywhere from one to five weeks; at their conclusion, the model auditor will render an opinion as to the reliability of the financial model. However, the primary purpose of the model audit is to define and reduce the amount of risk for the lenders relying on the model, and the model audit should not be primarily relied on as an insurance policy.

7. Security – common documents and roadblocks to obtaining security

7.1 Security panel

Project financing differs from other secured financing transactions in both the structure of the security to be provided and the level of assurance that the lenders will require over key aspects of the project. In a project financing transaction, security is provided at the project level by the project entity and by any direct or indirect parent of the project entity, including the sponsor, to give the lender full recourse in an enforcement scenario to the project assets and control of the project entity. As previously noted, depending on the creditworthiness of the project entity and the sponsor, security may also be required from other members of the corporate group. This not only means that lenders will obtain asset-level security and pledges of the equity interests of the project entity, but also ensures that the lenders have the right to perform key contracts such as the EPC contract and required government consents to sell the project assets or the project as a going concern in an enforcement scenario. The lenders will also want to ensure that their security is properly registered, so as to ensure that on enforcement, they have the first priority perfected security interest; and they will need to understand any roadblocks to taking security, such as stamp tax and powers of attorney.

While the security panel may seem complex, particularly with mining assets often located in foreign jurisdictions, having the right local counsel can be of great assistance in ensuring that the appropriate security is taken and that this is done correctly in accordance with local law.

Common guarantee and security documents in a project financing transaction are set out in Table 1.

Table 1. Common guarantee and security documents

Document	Purpose
Guarantee	While the project entity will generally be the direct borrower under the project loan facility, lenders will look for other forms of credit support with the ultimate view of being paid in full (whether from the direct borrower or otherwise). As such, lenders in project finance transactions look for third parties – generally, other members of the project entity's corporate group or joint venture partners – to be guarantors and to provide a guarantee, such that if the borrower does not pay, the lender can call on the guarantee as a form of repayment under the loan agreements. These guarantees are often secured guarantees, such that the guarantor also provides security over its assets to support the repayment obligations under the guarantee.
General security agreement (or other equivalent personal property security agreement)	Provides security over all personal property of the respective debtor (ie, the party which is providing the security, such as the project entity), which allows the lender recourse to the personal property of the debtor in an enforcement scenario. Personal property generally includes, but is not limited to, accounts receivable, inventory (including, once production has commenced, production held as inventory and the stockpile), and equipment of the debtor.
Mortgage/ debenture (or other equivalent real property security agreement) and other like security documentation over mining claims	Provides security over the real property of the debtor, as well as the mining claims. Having security over the mining claims will be a valuable piece of a lender's security panel as it looks to control the key rights to mine the project. The form of this document varies based on jurisdiction, as often mining rights may be classified as real or personal property, depending on the jurisdiction in which the project operates. Furthermore, depending on how the real property and mining rights are held by the debtor, consents may be required from key

continued on next page

Document	Purpose
	stakeholders, such as governments (eg, to the extent that surface or mineral rights are leased from governments) or First Nation groups that may have rights to the respective land.
Pledge of shares/equity interests	While personal property and real property security agreements give a lender security over the assets themselves, a pledge of shares/equity interests gives the lender the ability to take control of the project entity in an enforcement scenario by stepping into the shoes of its shareholder. Often, as previously discussed, the shareholder of the project entity is the sponsor, which will provide the pledge of equity interests as security for the obligations under its guarantee. A pledge of equity interests allows both for the ability to sell the project as a going concern and, as an alternative method, for a lender to realise on its collateral in an enforcement scenario where it may be impractical (eg, if a court order would be required to sell a particular asset) or cost prohibitive for the lender to enforce on the asset-level security.
Assignment of material contracts	Several material contracts which are required either to build or operate the project or to retain the interest of the project entity in the mining rights will be critical to the security panel of the lenders. Often, these agreements include the EPC contract, First Nations impact benefit agreements and other agreements that could not be easily replaced (eg, agreements for power or water in jurisdictions in which the resource is otherwise not readily available). To take security in these agreements, lenders will take an assignment of material contracts which creates a security interest in the material contracts and gives the lenders certain rights to these agreements in an enforcement scenario.

continued on next page

Document	Purpose
Tripartite agreement/ direct agreement	Part and parcel of the assignment of material contracts are direct agreements or tripartite agreements. While the assignment of material contracts creates a security interest in the material agreements, it does not necessarily give the lenders rights in the material contracts to the extent that counterparty consent is required under those agreements to assign the same in an enforcement scenario. As such, lenders often require a tripartite agreement or direct agreement, which is an agreement with both the project entity and the contract counterparty to provide lenders, among other things, with step-in rights under the material contracts, the rights to cure certain defaults and notice directly from the contract counterparty. While often the contract counterparty is a private contracting party, when the contract or consent in question is with a governmental authority or other key stakeholder group, this tripartite agreement can be extra sensitive, both from a relationship perspective and from a timing perspective, as the timing is largely in the control of the contract counterparty.

7.2 Stamp tax

When considering a security structure in a project financing transaction, one consideration that should not be overlooked is stamp tax. Stamp tax, which may also be called stamp duty, is a tax that is levied in some jurisdictions in order to have the document 'stamped'. While the stamping in most jurisdictions today is simply symbolic, the stamp tax still exists as essentially a transaction tax and varies widely in application from jurisdiction to jurisdiction.

While in most jurisdictions stamp tax is a nominal set amount, such as $10 per document, it is a factor that needs to be considered within the context of the security panel in a project financing transaction, as the fee will often need to be paid in advance to achieve an effective closing.

Furthermore, in some jurisdictions the stamp tax may be based on the principal amount secured under the relevant security document. This may make sense in jurisdictions where the amount secured is generally nominal, but may create perverse consequences in a project financing transaction where a percentage of the principal secured can lead to hundreds of thousands of dollars of stamp tax. Because of this risk, the existence and quantum of stamp tax should be ascertained at the beginning of any project financing transaction when the security panel is being developed.

7.3 Powers of attorney/signing formalities

Security documents in local jurisdictions often come with unique execution formalities that may be unfamiliar. Not having an appreciation for the execution formalities can cause undue delay in the closing of a project finance transaction while the parties work to comply with the formalities. Execution formalities should be addressed with local counsel in each jurisdiction at the outset of conversations about the security documents and panel.

One of the most common signing formalities is the requirement for the signing authority to execute the security documents in person before a notary public, which is often a requirement in civil law jurisdictions. This is often not possible or practicable due to the travel time and cost to the requisite jurisdiction for the signing authority, particularly with a moving closing date, and as such execution is often achieved by issuing a power of attorney in favour of a local law firm or other entity that routinely provides such services. Generally, for a power of attorney to be used in a local jurisdiction, it must be both notarised and legalised or apostilled. The mechanical requirements to legalise or apostille a power of attorney depend on both the country in which the power of attorney is being issued – particularly if such country is a party to the Hague Convention – and the country in which the power of attorney is being used, which will set its standard for authentication. In both instances, time must be factored in to ensure that the proper authentication steps are followed, so that the requisite power of attorney is in place for the closing of the project finance transaction.

7.4 Registrations/consents

Lenders will need to ensure that the security that they are taking – whether personal property security, real property security, a pledge of shares or otherwise – is properly registered in the appropriate jurisdictions, so that they have a better claim to the assets that they have security over in an enforcement scenario. Registration formalities vary widely based on the type of security that is taken and the jurisdiction in which the security is taken. While counsel in any jurisdiction will be relied on heavily to ensure that the registration is properly completed, the legal opinions of such counsel will also be expected to state that the security has been properly registered.

Taking security may also require various consents. Most commonly, to the extent that the mineral title is owned by a governmental authority and leased or otherwise licensed to the project entity, the consent of that governmental authority will be required in order both to charge the assets with the security and ultimately to realise on the security in an enforcement scenario. As previously noted, consent is generally required from contractual counterparties in order to assign material contracts. At the outset of a project finance transaction, in connection with the due diligence, a review should be done to determine what key rights or contracts require consent, and any third-party consents should be viewed as a matter of priority due to the time it may take to obtain such consents.

8. Working with local counsel

Local counsel play an important role in advising on the registration mechanics, security panel, stamp tax and possible unknown issues unique to the foreign jurisdiction. In addition, in jurisdictions that require government approvals or registrations, local counsel can help to execute on such deliverables. As detailed below, there are many possible issues that can delay a project financing in foreign jurisdictions. It is therefore imperative to hire local counsel familiar with mining and project finance transactions and to bring in local counsel as early as possible on these projects.

8.1 Security structures

Local counsel will be key in opining on the structure of the security panel in foreign jurisdictions, as local law may provide for different security structures with vastly different results on enforcement. Local counsel should identify the various options for a security structure, as well as the paths to enforcement, and the cost and ease of documentation. Involving local counsel in structuring decisions allows lead counsel to gain additional confidence that there will be a valid and enforceable security interest in the local jurisdiction.

8.2 Government approvals

Some jurisdictions require the approval of security interests by government authorities, particularly where the mining rights are owned by a governmental authority and contractually provided to a project entity. When working on financings in these jurisdictions, it is useful to engage local counsel for advice about the required governmental approvals necessary for the project and to review key documents which may not be in the language of the negotiations. There will often be several layers of bureaucracy involved to obtain these approvals and the necessary steps may be cumbersome. As such, these approvals may take months to obtain (or consent may not be possible, requiring lenders to rework the security panel). By engaging local counsel early, project entities can reduce the risk that these security interests are not approved, ensuring that the project financing stays on time.

8.3 Searches and certificates of good standing

It is also important to discuss searches and certificates of good standing with local counsel, which identify any encumbrances on the assets or issues that may exist for the project company, such as material litigation. In foreign jurisdictions, these documents can take several weeks or months to obtain and may require multiple searches, but are key to understanding existing issues. Early advice from local counsel will help to keep the advisers informed about these possible search timelines – which may be longer than one would expect – and oddities such as the requirement to physically attend offices to conduct searches.

8.4 Registrations and fees

Local counsel can also advise on the process and timelines to register and guarantee security documents. In some jurisdictions such as Brazil, documents must be presented for registration in person at the local state registry office. In others, such as Peru, the documents must be filed or submitted for recording with the applicable public registry. These requirements add additional time and cost constraints, which can also jeopardise closing if not adequately handled. By retaining local counsel, advisers can stay informed of these additional components; and where the local jurisdiction requires in person registration, local counsel can help in physically attending the registry to facilitate execution of the documents.

8.5 Language of documents

Local counsel can also help to navigate possible language-related issues through the project financing transaction. First, local counsel can advise on both how and in what language the documents should be negotiated, and on any local language requirements. Where documents are negotiated primarily in one language, they are often still drafted in the parties' native language. In such cases it is essential to obtain a translation of the draft for the negotiations on a non-reliance basis and on the premise that the original language document governs. If, in contrast, the documents are not negotiated in English, local counsel can also be relied on to ensure that the documents reflect the deal terms.

8.6 Other

Local counsel can also advise on other non-legal considerations. For example, it is advisable to ask local counsel early on for a list of upcoming public holidays and to expect that local counsel will be unavailable during those public holidays. Another consideration is to recognise and take into account the time zone differences. Depending on the location of the foreign jurisdiction, there may only be a few hours each day in which all parties are available.

9. Conclusion

Project financing in the mining sector requires a number of moving parts to come together to result in a successful and profitable mining operation. By working with advisers experienced in project finance in the mining sector, even the most complex multi-jurisdictional projects can be financed so as to allow for the risks of the project lenders to be managed while still allowing for construction of the project on time and on budget.

Sale and purchase of mineral projects

Darrell W Podowski
Jennifer Poirier
Cassels Brock & Blackwell LLP
Felipe Allende
Bofill Mir & Álvarez Jana Abogados

1. Introduction

This chapter provides an overview of the sale and purchase of mineral projects, from grassroots exploration projects to operating mines, and includes a summary of key factors to be considered when determining the appropriate structure for such transactions, as well as a summary of the key features of these transaction structures. It provides a high-level overview of pertinent considerations as applicable in Canada, although many of these concepts will also be applicable in other jurisdictions. In general, it does not touch on the application of these considerations to specific jurisdictions; however, to the extent that illustrative examples may be useful, specific examples of relevant laws and regulations applicable in Canada, and in some cases Chile, have been included.

2. Factors to consider in determining transaction structure

Early in the negotiation process, the parties must decide whether the transaction will be structured as a direct purchase and sale of the assets comprising the mineral project or as a purchase and sale of the shares of the company (eg, a special purpose vehicle (SPV)) that holds the assets comprising the mineral project. As might be expected, the purchaser and the vendor may disagree on the preferred approach. A range of factors may be considered in determining whether to structure a mining transaction as an asset acquisition or a share purchase. There follows a brief discussion of a select list of these factors. No one factor is necessarily determinative, as each transaction must be evaluated on the basis of its unique facts. That said, the factors considered in sections 2.1, 2.2 and 2.3 are often particularly influential in determining transaction structure. In addition, the relative bargaining power of the parties to the transaction may influence the weight given to any particular factor. For instance, in a transaction involving a senior company and a junior company, the factor or factors that are most important to the senior company may prevail, due to the unequal negotiating strength of the parties.

2.1 Stage of the project

(a) *Early stage*

For a transaction involving an early stage mineral project (grassroots exploration), the project will likely have simple mineral tenure, minimal technical data and minimal permits, and will likely be owned by a junior company.

Where this structure is present, all other factors aside, the transaction will often lend itself to being structured as an asset purchase in the form of either an option or a joint venture (as is often the case where the transaction involves the purchase of less than 100% of the assets comprising the mineral project), or a purchase of 100% of the assets comprising the mineral project (see sections 4.1 and 4.2).

(b) *Advanced exploration*

Many more considerations come into play if the transaction involves an advanced exploration project. For instance, because the parties will likely be dealing with more advanced mineral tenure, this could give rise to consent and transfer issues (see section 2.4(h)). In addition, for an advanced mineral exploration project, there will likely be much more technical data, potentially giving rise to IP, disclosure and confidentiality issues; and a greater number of permits and licences, potentially giving rise to further consent and assignment issues. There could also be environmental issues, such as environmental liabilities associated with exploration activities conducted on the project. The vendor may also have commenced consultation with local communities and/or indigenous groups, and the negotiation of related agreements. An advanced exploration project could be owned by a junior, mid-level or senior company.

For this stage of the project, the transaction could be structured as an option or joint venture, or as a purchase of 100% of the assets comprising the mineral project. On the other hand, the transaction could be structured as a share acquisition, whereby the purchaser instead acquires ownership of the entity that holds the assets comprising the mineral project (often an SPV).

(c) *Development/near production*

For development or near production projects, the project will likely be of much higher value, so greater focus will be placed on valuation of the project and determining appropriate consideration and the preferred tax treatment for the acquisition. In addition, because the project will likely have mature mineral tenure, and many permits and licences, governmental and other third-party consents to transfer may be required; this could lean in favour of structuring the transaction as a share acquisition. The presence of a large amount of technical data and intellectual property and advanced arrangements, and negotiations with local communities and/or indigenous groups, may also support structuring the transaction as a share purchase, as there could be issues with transferability and goodwill. On the other hand, if there are many contractual obligations and liabilities that the purchaser does not want to inherit, this factor could lend itself to structuring the transaction as an asset purchase. Another relevant consideration is whether there will be severance and/or retention issues with regard to employees and management. A development/near production project will likely be owned by a mid-tier or senior company.

If a number of the factors described above are present, the transaction may be more likely structured as a share purchase, pursuant to which the purchaser will acquire ownership of the entity that holds the assets comprising the mineral project (often an SPV), or of the whole parent company (which may be a private or publicly listed company), if the mineral project is that company's only material asset.

(d) Producing

As may be expected, for a producing project, there will be the greatest number of considerations compared to all other project stages. A producing project will have a much higher value, so valuation and tax structuring will be of utmost importance. Mineral tenure will be mature and there will likely be a number of permits and licences – both factors which may necessitate obtaining governmental and other third-party consents to transfer. In addition, there will likely be:

- a large amount of technical data and intellectual property;
- advanced arrangements and negotiations with local communities and/or indigenous groups;
- a larger number of contractual arrangements (including possibly hedging or streaming arrangements); and
- a significant number of employees.

A producing project may also have significant liabilities (including environmental) and reclamation obligations (eg, bonds posted). This type of mineral project will most likely be owned by a mid-tier or senior company.

This type of transaction is more likely to be structured as a share acquisition of the operating company holding and operating the assets comprising the mineral project or of the whole parent company (which may be a private or publicly listed company), if the mineral project is that company's only material asset.

2.2 Tax

Quite often, the vendor will prefer to structure the transaction as a share purchase, due to the favourable tax consequences for the vendor that can result from this type of transaction. In particular, a share transaction can result in capital gains treatment for the vendor, rather than an income inclusion.

A share transaction may also be favoured by the purchaser, as it may allow the purchaser to take advantage of favourable tax attributes of the vendor, such as tax loss carry forwards. Generally, these tax attributes will not benefit the purchaser if the transaction is structured as an asset purchase. However, if favourable tax attributes are not present, it may be more advantageous for the purchaser to purchase the assets, because the tax basis for the assets will be equal to the acquisition purchase price, possibly resulting in future tax savings. In addition, within the context of an asset purchase, the purchaser can negotiate the allocation of the purchase price among the resource property, inventory, depreciable and non-depreciable capital property; any amounts allocated to the resource property, inventory or depreciable property may be more tax effective for the purchaser than those allocated to other property. Another important consideration is that sales and land transfer taxes may be payable in connection with an asset purchase which would not be payable in a share purchase.

For many years, the above considerations were reflected in Chilean law, which provided that a transaction structured as a share purchase was in most cases subject to a relatively low capital gain tax which was equivalent to the corporate income tax rate (24% in 2016). However, the 2014 Tax Reform modified the treatment of capital gains, so that since 2017, capital gains from Chilean companies other than stock companies with a certain market volume are now subject to a 35% capital gain tax in the hands of foreign investors. On the other hand, a transaction structured as an asset purchase will present some disadvantages for the vendor, such as the following:

- Most likely, the assets will already be depreciated and therefore there will be a low tax value which the vendor will be unable to deduct from the applicable tax basis after receiving payment for the assets;
- In many cases, a transaction structured as an asset purchase will be subject to value added tax, equivalent in Chile to 19% of the price of the affected asset; and
- Since mining companies are intensive in their use of fixed assets, many of them are in a loss position which can be carried forward if the transaction is structured as a share purchase.

There may also be tax advantages to structuring the transaction using a subsidiary that is resident in a particular jurisdiction. For instance, in Canada, there may be benefits if a foreign purchaser uses a Canadian subsidiary to complete a share purchase transaction of a Canadian target. A post-acquisition merger or amalgamation of the target and Canadian subsidiary may permit the deduction of interest on financing for the transaction against the Canadian business income and may assist in repatriating funds to the foreign purchaser without triggering withholding tax. Where the transaction is structured as an asset acquisition, pursuant to which a foreign purchaser acquires Canadian business operations, that foreign purchaser will be responsible for the liabilities of the Canadian business which are assumed by the purchaser and for Canadian tax on income generated from such business. If the foreign purchaser uses a Canadian subsidiary to acquire the assets instead, the Canadian subsidiary, rather than the foreign purchaser, will be responsible for the assumed liabilities and Canadian tax on the income generated from the Canadian business.

In Chile, there is a 35% non-resident withholding tax on dividends or remittances of profits to non-Chilean shareholders. The corporate income tax paid by the Chilean company can be credited against the withholding tax. However, the amount of the tax credit depends on the system under which the Chilean company is taxed. Companies taxed under the semi-integrated system are subject to a 27% corporate income tax and when there is a dividend distribution abroad, withholding tax is triggered. Some 65% of the corporate income tax can be credited against the withholding tax, which means that foreign investors bear a 44.45% total tax burden. However, a foreign investor based in a country which has entered into a double tax treaty with Chile is entitled to fully credit the corporate income tax against the withholding tax, resulting in a 35% total tax burden. Meanwhile, companies taxed under the attributed system are subject to a 25% corporate income tax. In contrast to the semi-integrated system, under the attributed system, the 35% withholding tax must be paid each year on annual taxable income, regardless of when dividends are paid or remitted abroad. However, foreign investors taxed under the attributed system are entitled to credit the entire corporate income tax paid by the Chilean company. As a consequence, foreign investors under the attributed system bear a 35% total tax burden.

As noted above, since 2017, transactions structured as a share purchase have been subject to a 35% income tax in the hands of foreign investors. A provisional tax of 10% must be withheld by the buyer

2.3 Liabilities

The vendor may be inclined to have the transaction structured as a share purchase, because this allows the vendor to step away from the business, as the purchaser assumes all liabilities and contractual obligations of the target. In a similar vein, the purchaser may prefer to structure the transaction as an asset purchase, as this type of transaction provides the purchaser with flexibility in picking and choosing the exact assets to be acquired and, to a certain extent, the liabilities that the purchaser will assume. Of course, not all liabilities can be avoided, as certain liabilities may be assumed by the purchaser by operation of law, such as environmental liabilities that run with the land. Accordingly, if there are material liabilities, a purchaser will likely prefer an asset deal over a share transaction.

2.4 Other factors

(a) *Foreign investment issues*

Depending on the jurisdictions at issue, the vendor's jurisdiction may have rules regarding foreign investment that could affect the transaction structure. For instance, under the Investment Canada Act, a non-Canadian that acquires direct control of a Canadian business whose value is above a certain monetary threshold must apply to the Investment Review Division for review of the transaction and must be approved by the minister of innovation, science and economic development. A non-Canadian may be deemed to acquire control of a Canadian business where it acquires more than one-third, but less than a majority, of the voting shares of the corporation, and will acquire control where it acquires a majority of the voting shares of the corporation or all or substantially all assets of the business.

If the transaction is below a certain monetary threshold, the transaction is not reviewable, but the non-Canadian purchaser is nevertheless required to provide the government with notice of the transaction; notification can be made in advance or within 30 days of closing. There are different monetary thresholds depending on the jurisdiction out of which the purchaser or vendor is ultimately controlled, or whether the purchaser is a state-owned enterprise. For 2018, the monetary thresholds for review of direct investments by World Trade Organization (WTO) member investors or trade agreement investors are as follows:

- Where the purchaser or vendor is ultimately controlled out of a WTO member country and the purchaser is not a state-owned enterprise, the threshold will be met where the Canadian business has an 'enterprise value' of C$1 billion;
- Where the purchaser or vendor is ultimately controlled out of a country with which Canada has a private sector trade agreement[1] and the purchaser is not a state-owned enterprise, the threshold will be met where the Canadian business has an enterprise value of C$1.5 billion; and
- For purchasers which are state-owned enterprises operated by WTO member countries, the threshold will be met where the Canadian business has an asset value of C$398 million, based on the book value of the assets.

'Enterprise value' for public and private entities is determined as follows:

- For acquisitions of control of publicly traded entities, the enterprise

1 For example, the Canada-European Union Comprehensive Economic and Trade Agreement Implementation Act, the North American Free Trade Agreement, etc. A full list of private sector trade agreements is set out under Section 14.11(6) of the Investment Canada Act and the accompanying schedule.

value of the assets of the Canadian business is equal to the market capitalisation of the entity plus liabilities, minus cash and cash equivalents; and
- For acquisitions of control of private companies and for asset acquisitions, the enterprise value is the purchase price, plus liabilities, minus cash and cash equivalents.

Pursuant to Canada's international commitments, indirect acquisitions of a Canadian business by or from a WTO investor are not reviewable, unless they raise national security issues or involve the acquisition of a cultural business.

For purchases by non-Canadian investors which are not ultimately controlled out of a WTO member country, the threshold is significantly lower, at C$5 million for direct transactions and C$50 million for indirect acquisitions. These thresholds also apply to all investments by non-Canadians to acquire control of a Canadian business that is a cultural business as described in Section 14.1(6) of the Investment Canada Act.

If the transaction is reviewable under the Investment Canada Act, the initial review period is 45 days, which can be unilaterally extended by the minister for a further 30 days. Further extensions of the review period require the consent of both the purchaser and the minister. To obtain approval, the minister must be satisfied that the acquisition is likely to be of "net benefit to Canada".

In addition to the above types of review, the government may review any proposed acquisition of a Canadian business (including a minority interest) or acquisition of a business with some Canadian operations, where it has reasonable grounds to believe that such proposed acquisition may be "injurious to national security".

In the context of Chilean law, it is not necessary to create a company under domestic law or to acquire an interest in a mining company in order to hold an exploration and/or mineral title. Further, with regard to mineral title, Article 22 of the Chilean Mining Code establishes that any person is authorised to request or to acquire mining concessions; while it states specific exceptions to this general rule, the award of mining concessions to a foreign company is not one of them. Therefore, any mining concession or mining company can be wholly owned by a foreign party.

However, it is nonetheless recommended to incorporate a domestic company to hold the mining concession for practical reasons, such as the payment of application fees and representation before the courts. Moreover, the establishment of a local entity affords the owner greater flexibility with regard to tax matters – for example, it can reinvest profits locally prior to paying any withholding tax for the distribution of dividends abroad.

That said, Article 7 of Decree Law 1.939 establishes that nationals of neighbouring countries are prohibited from acquiring the domain, possession or even tenancy of real property (including mining concessions) situated wholly or partly in certain areas of the country. This prohibition is applicable to both individuals and legal entities established in such neighbouring countries, and legal entities whose capital is at least 40% owned by nationals of such countries or whose administration and effective control are exercised by nationals of such countries. Specific exceptions can be made to this law.

(b) Competition issues

In addition to complying with any foreign investment legislation, the parties must be mindful of competition merger control laws that may apply to the transaction. In Canada, under the Competition Act (Canada), the commissioner of competition can challenge a transaction that he believes could substantially prevent or lessen competition. The Competition Act requires pre-notification of transactions that exceed certain thresholds. For instance, pre-merger notification is required for transactions where the parties (including their affiliates) have Canadian assets or revenues from sales in, from or into Canada that exceed C$400 million, and where the assets being acquired in Canada or the revenues from sales in or from Canada generated by such assets exceed C$92 million.

If the transaction is a notifiable transaction under the Competition Act, the parties must comply with statutory filing requirements and a waiting period must pass before closing the transaction. If the transaction raises significant competition-related concerns, the commissioner of competition may request additional information; this request will extend the waiting period for another 30 days from the date on which the parties provide the requested additional information. If the commissioner of competition concludes that the transaction raises no significant competitive concerns, he may provide an advance ruling certificate or a 'no-action' letter indicating that the transaction will not be challenged under the Competition Act.

(c) Percentage of the project that the vendor owns

If the vendor does not own 100% of the assets comprising the mineral project (eg, where the property is held through a joint venture), or if the vendor is currently still earning its interest in the property (eg, through an option agreement), the transaction between the vendor and the purchaser is also likely to be structured as a joint venture or option agreement, as applicable.

If the vendor is the direct holder of 100% of the assets comprising the mineral project, this can allow for greater flexibility in determining transaction structure. The purchaser could seek to acquire some or all of the assets comprising the mineral project, and the acquisition could proceed by asset or share purchase.

(d) Vendor's only assets

If the assets comprising the mineral project are the only assets of the vendor, the purchaser may opt simply to purchase the entire company, in which case the transaction will be structured as a share purchase. However, if the vendor holds multiple assets, an asset acquisition may be the preferred approach, as it will allow the parties to precisely parcel off those assets being acquired from the assets being retained by the vendor.

(e) ***Nature of consideration***

Another relevant factor is the nature of the consideration to be offered by the purchaser. If the purchaser is prepared to offer all cash, this allows for greater flexibility in transaction structure. However, if the purchaser prefers to issue shares or offer some of its own assets as consideration, this can not only give rise to tax and securities laws issues that must be analysed, but also lead to less flexibility in terms of deal structure.

(f) ***Type/size of vendor***

If the vendor is a senior company, the manner in which the asset is held could dictate the transaction structure that is implemented. For instance, if the asset is already held through an SPV, the transaction will likely be structured as a share purchase of that SPV. In addition, senior companies may be more favourable to accepting shares as consideration – a factor which lends itself to structuring the transaction as a share purchase. On the other hand, if the vendor is a junior company, it will be more likely to require cash consideration, as compared to shares, as junior companies often need the cash to be put into the ground in order to advance the exploration of the mineral project.

(g) ***Shareholder base***

If the vendor is a private closely held company, this may favour a simple share purchase transaction completed via execution of a share purchase agreement (see section 5.1). However, for private companies that have a large shareholder base, and for public companies, the type of share purchase will be more complex and is likely to take the form of a plan or scheme of arrangement, amalgamation or takeover bid. In addition, if the shareholder base of a Canadian vendor includes US shareholders and the transaction will involve the issuance of securities as consideration, the transaction may likely be structured as a plan or scheme of arrangement, due to the availability of certain registration exemptions for share issuances, under US securities laws (see section 5.2).

Even if the transaction is structured as an asset purchase, this will not necessarily dispense with the requirement to obtain shareholder approval, as the approval of shareholders will likely be required for asset purchases that involve a sale of all or substantially all assets of the company, which often requires shareholder approval at a level greater than a simple majority.

(h) Mineral tenure, permits and material contracts assignable

In an asset purchase, various assets (eg, mineral tenure, permits, material contracts) must be transferred or assigned from the vendor to the purchaser. As a result, it is necessary to determine whether the assets can in fact be transferred or assigned. In many jurisdictions of Canada, mineral claims and mineral leases are transferable, though the transfer must comply with certain legislative requirements. A general requirement for such transfers is that they be executed in writing and signed by the claim holder or lease holder. However, depending on the jurisdiction, there may also be stricter rules that must be complied with. For instance, in Nova Scotia, transfers of certain mineral rights require the consent of the minister of natural resources (pursuant to the Mineral Resources Act (Nova Scotia)). It is also necessary to consider whether the consent of any governmental authority or third party is required prior to the transfer or assignment of any permits or material contracts. Consent of the applicable First Nations groups may also be required in Canada.

The situation is very similar under Chilean law, as both types of mining concessions (exploration concessions and extraction concessions) are transferable. However, such transfer must comply with certain legal requirements, including that the transfer be executed in writing, by means of a Chilean public deed. Additionally, registration of the transfer in the corresponding mining register is legally required. The administrative body in charge of the mining register is the Mining Registrar's Office, which is governed by the Mining Code and the Mining Code Regulation. The Mining Registrar's Office maintains several registers, including the Register of Discoveries, the Register of Properties, the Register of Mortgages and Encumbrances and the Shareholders Register.

In Chile, any initial petition for an exploration concession or for an exploitation concession, and any transfer of such petitions, must be registered in the Register of Discoveries. The judicial resolution granting an exploration concession and approving any transfer of such concession must also be registered in the Register of Discoveries. The judicial resolution granting an exploitation concession and approving any transfer of such concession must be registered in the Property Register.

Furthermore, mortgages, usufructs and encumbrances over the mining concession must be registered in the Mortgages and Encumbrances Register and, with respect to the two types of special mining entities (*Sociedad Legal Minera* and *Sociedad Contractual Minera*), the list of shareholders. In addition, any transfers and pledges on such shares must be registered in the Shareholders Register. If registration is required, failure to register may result in annulment of the corresponding right.

If a significant number of consents are required, or if one or more consents will likely be difficult or time consuming to obtain, this may tilt in favour of structuring the transaction as a share purchase. Nevertheless, even if the parties proceed with an asset purchase, they will still need to be mindful of any change of control provisions (eg, in material contracts) that could trigger the requirement for consent or advance notice of the transaction (see section 4.1).

(i) *Goodwill that the purchaser is hoping to maintain*

If there is any desire to retain goodwill, intellectual property, technical or geological knowledge or employees, a share purchase is the most effective structure for accomplishing this.

(j) *Timing*

Timing preferences could influence transaction structure, as it may be possible to sidestep the need to obtain government or other third-party consents required to transfer an asset if the transaction is structured as a share purchase.

3. Due diligence considerations

In addition to determining transaction structure, another important task that commences early in the transaction process is due diligence. Once again, the stage of the project is a significant factor in determining the scope of the due diligence to be conducted. If the project is early stage, due diligence may be fairly minimal. However, as the project progresses to later stages (advanced stage and beyond), the due diligence becomes increasingly more complex, particularly with regard to mineral tenure, technical data, permitting and licensing, environmental matters and relationships with local communities and indigenous groups.

Another related consideration is whether the purchaser will request that a title opinion be prepared in connection with the transaction. A title opinion is for the purchaser's protection and can be particularly important if the purchaser is conducting a concurrent financing to raise funds to finance the acquisition. The title opinion could reveal significant issues, such as the fact that the vendor does not actually have clear or clean title, or that the asset is near the end of life of the mineral tenure. For this reason, purchasers should not only request to review previous title opinions as part of their due diligence, but also request to see a draft title opinion well in advance of closing, if a new title opinion is required in connection with the transaction.

4. Asset acquisitions

Once the parties have determined that the transaction will be structured as an asset acquisition rather than a share acquisition, it is then necessary to consider which type of asset acquisition will be implemented. Below is a description of two different types of asset acquisition:

- acquisitions of the entire asset; and
- options and joint ventures.

4.1 Acquisitions of entire asset

If the parties determine that the preferred structure is an acquisition of the entire asset (eg, 100% of the assets comprising the mineral project), the transaction will involve the purchase of just these assets, rather than the acquisition of the entity holding such asset. The parties will negotiate the terms of an asset purchase agreement that may or may not contain representations and warranties regarding the asset, such as:

- absence of environmental liabilities;
- ownership of the mineral project;
- existence of material mineral licences or permits in good standing; and
- operations in respect of the mineral project having been conducted in material compliance with laws.

For sales that are 'as is where is', however, these types of representations and warranties will not be present in the agreement.

An asset acquisition can be a very quick process, but a few factors could significantly increase the length of time to complete the transaction. Because the asset acquisition involves the transfer of assets from the vendor to the purchaser, the consent of governmental authorities and other third parties may be required prior to completion of the transaction. In addition, consents may be required to assign the material contracts related to the asset. For this reason, the filings and consents required for an asset acquisition could be more onerous than those required for a share purchase.

Another factor that could increase the time to complete the transaction is the requirement for shareholder approval. Corporate laws applicable to the vendor may require shareholder approval, even if the transaction is structured as an asset acquisition. For example, in Canada, corporate statutes (eg, the Canada Business Corporations Act) require that a sale of all or substantially all assets of a company (eg, the sale of the only asset of a company) be approved by at least two-thirds of the company's voting shares.

The Chilean corporate statutes are very similar in this regard: in most cases, the sale of more than 50% of the assets of a company must be approved by at least two-thirds of the company's voting shares.

4.2 Options and joint ventures

(a) *Option agreements*[2]

An option agreement is an agreement between the owner/holder of a mining right (referred to in this section as the 'optionor') and a purchaser (referred to in

2 Refer to the 'Option agreements in the mining industry' chapter by Shea Small, Roger Taplin and Adam D Wanke for a more detailed discussion of this topic.

this section as the 'optionee'), pursuant to which the optionee is granted the option to acquire by transfer all or a portion of the mineral or mining rights upon the fulfilment of specified financial and work performance conditions. Option agreements allocate and mitigate exploration risk, because while the optionee is required to incur certain expenses in order to exercise the option, if the optionee determines that it is not satisfied with the results, it can instead elect not to proceed with the option and discontinue any further expenditures on the property (subject to any requirements in the option agreement to mandatorily fund certain amounts – referred to as 'committed' expenditures – or to continue to fund certain expenses post-termination). Option agreements are often relied upon by junior companies to get value from a mining project that they are not capable of exploring, developing and financing on their own. They also serve as a means for mid-tier and senior companies to get access to potential properties in areas or countries in which they may have no operations or experience.

As briefly mentioned above, expenditure requirements can be committed amounts, in that the optionee is required to pay them regardless of whether it chooses to exercise the option. Expenditures can also be optional, but mandatory, in the sense that in order to exercise the option, the optionee must fulfil the expenditure obligations; however, if the optionee chooses not to fund such obligations, the optionor has no claim against the optionee for failing to fund. In this case, the decision not to fund will terminate the option.

There is also some flexibility in the form that the option conditions can take. For instance, the option conditions could require only cash payments, work and qualifying expenditures (usually conducted within a set timeframe), the delivery of a report or a study, such as a scoping study, or any combination of these.

The operator of work programmes on the property may or may not be the optionee. If title to the property is not transferred to the optionee until later in the option, the optionee will need to ensure that it can conduct operations on the property if it is not the title holder.

Upon exercise of the option, title to the mining rights acquired may be transferred and assigned to and registered in the name of the optionee if title was not transferred at the start of the option, or the optionee may be granted shares in the company (eg, an SPV) that holds the mineral project. At some point, the optionor and optionee will enter into a joint venture agreement if the acquired interest is less than 100% (see section 4.2(b)).

(b) Joint venture agreements[3]

While a precise definition of 'joint venture' does not exist, a joint venture is generally an arrangement among parties that retains the individuality of the

3 Refer to the 'Joint ventures' chapter by Stuart MacGregor and Marie-Christine Moon for a more detailed discussion of this topic.

parties and that relates to a single project or venture. Two main types of joint ventures are:

- contractual joint ventures, in which each participant holds an undivided direct interest in the mineral property and assets, as tenants in common, expressed as a percentage; and
- corporate joint ventures, which are not in fact true joint ventures, as they involve the incorporation of an entity to own and operate the project.

In a corporate joint venture:

- each party to the joint venture owns shares in the corporation in proportion to its percentage interest in the venture;
- a unanimous shareholders' agreement governs the relationship between the owners; and
- the corporation holds the title to the mineral property and assets.

Joint ventures can arise out of:

- co-ownership resulting from exercise of an option pursuant to an option agreement;
- a sale of less than a 100% interest in a property; or
- an agreement whereby one party contributes the property and another party contributes other assets to the venture, such as other property, financing or expertise.

One benefit of a corporate joint venture is that the existence of a separate corporate entity can provide some protection for unwanted liabilities. This is because the corporate entity, and not the joint venture partners, will (except in limited circumstances) be on the hook for the actions and liabilities of the corporate joint venture. In addition, it allows for a separation between management and ownership. That said, the joint venture partners or their respective representatives will likely become directors and officers of the corporate joint venture, which itself can attract certain duties, obligations and liabilities.

A joint venture is generally governed by a management committee (or the board of directors of the corporation in the case of a corporate joint venture), with each joint venture party having representation on the management committee or board proportionate to its percentage interest in the venture. The manager/operator of the project is usually one of the joint venturers, although it may also be a contracted third party. The manager/operator is generally entitled to reimbursement by the joint venture partners for expenditures and to a fee for its services.

The manager/operator will compile and present programmes and budgets for operations or the property to the management committee or the board, as applicable. The management committee or the board, as applicable, will generally either approve, or amend and approve, the proposed programme and budget. During the earn-in stage of such option transaction, the optionee will typically be responsible for 100% of all programmes and budgets on the property (ie, it 'carries' the optionor). Thereafter, where the optionee earns less than a 100% interest in the property, once a programme and budget is approved by the management committee or the board (as applicable), the joint venture parties must determine whether they will elect to participate in the programme and budget. If the decision is to participate, the joint venture party must fund its *pro rata* portion of the programme and budget within the required timeframe. If a joint venture party elects not to fund, or elects to fund but then fails to fund within the required timeframe, that non-contributing party will likely suffer dilution or forfeiture of its interest, depending on the negotiated terms. To avoid having small percentage ownership interests, often a party's interest will convert to a royalty if it dilutes below a certain threshold (eg, 10%).

Upon commencement of commercial production, the joint venture agreement will often contemplate distributing product in kind to the venture parties in accordance with their joint venture interests. One party may also have the other party or the manager/operator sell the product on its behalf and account back.

5. Share acquisitions

If the parties determine that a friendly acquisition of the entire company is the preferred approach, then additional factors will need to be considered to determine the precise transaction structure that is best suited to the acquisition. Below is a description of the following types of share acquisitions:

- private company share acquisitions; and
- public company share acquisitions.

5.1 Private company share acquisition

If the target is a private company, the simplest approach may be to structure the transaction as a straight share purchase. The parties will negotiate the terms of a share purchase agreement that will contain representations and warranties regarding the target and the conduct of the target's business, including those with regard to due incorporation, authority to carry on business, ownership of assets, compliance with laws, absence of environmental liabilities, indebtedness and ownership of the mineral project. Because the entire company is being acquired, there may be less need to obtain governmental and third-party consents. However, it will be necessary to review the terms of permits, authorisations, licences, contracts and so on in order to determine whether the proposed structure would constitute a change of control under any such documents. For instance, often the acquisition by a person, or group of persons acting jointly or in concert, of more than 50% of a company's shares will constitute a change of control. If the transaction constitutes a change of control, the target may be required to obtain prior consent from, or at the very least provide prior notice of the transaction to, a third party or governmental authority.

In addition, even if the company is private, if the shareholder base is widespread, it may be more efficient to structure the transaction using one of the statutory procedures described in more detail under section 5.2(b), such as a statutory amalgamation, merger or plan or scheme of arrangement.

5.2 Public company share acquisitions

If the target is widely held or is a public company, the share acquisition can generally be structured in one of three ways:

- a takeover bid;
- a statutory amalgamation or merger; or
- a statutory plan or scheme of arrangement.

Each of these is described in more detail below and may or may not be available in the jurisdiction of incorporation of the target. The summary below applies in the context of a friendly transaction in a Canadian context; different considerations (which are not discussed here) will apply if the transaction is hostile or governed by legislation in another jurisdiction.

(a) Takeover bids

A takeover bid is an offer made directly to security holders to purchase 100% of the securities of the target at a pre-determined price. The purchaser (referred to in this section as the 'offeror') must offer identical consideration to all target security holders of the same class. In a friendly takeover bid, often the offeror and the target will enter into a support agreement whereby the target will agree, subject to the terms of the support agreement, to support the takeover bid; the offeror may also attempt to get the agreement of a larger shareholder to support the bid and 'lock it up' (subject to any regulatory restrictions that may be applicable). The bid is generally made by way of a circular sent to target security holders; the circular must describe all terms and conditions of the offer, including the securities being offered as part of the consideration, if applicable. The bid must be open for security holder acceptance for a certain period and may be subject to minimum tender requirements. In Canada, for instance, pursuant to National Instrument 62-104 – Take-Over Bids and Issuer Bids, the bid must be open for at least 105 days, subject to certain limited exemptions. Bids in Canada are also subject to a 50% minimum tender requirement; once that minimum tender requirement is satisfied, the bid must be extended for an additional 10 days.

If 100% of the securities are not tendered to the bid (which is usually the case), there may be a mechanism to acquire the remaining securities if a certain percentage of securities are tendered to the bid. Under Canadian corporate law (eg, the Canada Business Corporations Act), for example, if 90% or more security holders tender to the bid (and certain other requirements are met), the offeror can easily force the remainder to tender their securities. If less than 90% but more than 66⅔% of the securities are tendered, the offeror can acquire the remaining securities by completing a second-stage squeeze-out transaction. Often this second-stage transaction takes the form of an amalgamation or an arrangement which is approved by 66⅔% of shareholders of the target. The consideration under the second-stage transaction must be identical to the consideration offered under the takeover bid.

An advantage of a takeover bid is that these transactions do not require a judicial determination of fairness. In addition, takeover bids can be pursued unilaterally (ie, without the support of the target), with the effect that the offeror, rather than the target, can maintain maximum control of the process. Also, either cash or shares can be offered as consideration. The offeror can also flexibly respond to competing bids by increasing or varying the terms of the consideration.

However, in Canada, as most takeover bids are required to remain open for a minimum of 105 days, takeover bids can take significant time to complete. Additionally, bids are subject to the 50% minimum tender requirement, with the result that bids may be unsuccessful if the tender threshold is not met. Other potential disadvantages are the fact that the bid must offer identical consideration to all security holders and the target is still in play during the bid, meaning that a competing bid could emerge.

(b) Statutory schemes

Plans of arrangement: A plan of arrangement under Canadian corporate law (similar to a scheme of arrangement under Australian law) is a merger affected by obtaining court and security holder approval. The purchaser and the vendor will negotiate the terms of an arrangement agreement that will contain certain representations and warranties regarding the target and the conduct of the target's business. The agreement will also contain certain covenants regarding conduct of business in the ordinary course and covenants regarding completion of the transaction. The scope of the representations and warranties and the covenants may or may not be reciprocal between the parties, depending on the nature of the transaction. For instance, in a transaction where a senior company is seeking to acquire a junior company, the senior company may provide much more limited representations and warranties and covenants to the junior company. In addition, the agreement will contain certain conditions precedent to completion of the transaction, and will likely contain certain non-solicitation provisions which seek to prevent (or at least reduce the likelihood of) competing offers. The parties may also negotiate the payment of a termination fee which becomes payable in the event that the non-solicitation provisions are breached.

Generally speaking, following negotiation of the arrangement agreement, the parties will apply to court to obtain an interim order. The interim order will set out the requirements for the shareholders' meeting, including the classes of securities that have the right to vote separately as a class, the shareholder approval percentage required (often this is $66^{2}/3\%$), and the rights of shareholders to dissent to the transaction. Dissent rights provide shareholders with the ability to elect not to participate in the transaction and instead be paid an amount equal to the fair value of their shares; shareholders must exercise dissent rights in strict compliance with the provisions set out in the applicable corporate statute, as modified by the plan of arrangement and the interim order. Once the interim order has been obtained, proxy materials describing the nature of the transaction are sent to security holders. Once the shareholders have approved the transaction, the parties must apply to court to obtain a final order. If the court determines that the transaction is fair to the shareholders, it will grant the final order. Once shareholder and court approval has been obtained, subject to the satisfaction of other closing conditions (including receipt of any other regulatory approvals), the parties will close the plan of arrangement. From start to finish, a plan of arrangement generally takes 60 to 90 days to complete.

Plans of arrangement are the most common form of statutory transaction in Canada. The advantages to this type of transaction are that they entail only a single step to acquire 100% of the shares of a target and they are the most flexible, such as for dealing with outstanding options and warrants of the target. For instance, under the terms of the arrangement, options and warrants that are outstanding immediately prior to the effective time of the arrangement can be adjusted in accordance with their terms, such that they become exercisable to acquire the same number of shares that the holder would have been entitled to acquire had it exercised the options or warrants prior to the effective time. They also allow for tax planning. The fact that the transaction requires court approval also ensures a level of fairness for shareholders and provides a forum for shareholders to express their disagreement with the transaction.

In addition, if the transaction involves a US purchaser and it is proposed to be structured as a friendly share purchase (as compared to a hostile transaction), whereby shares of the US purchaser will be exchanged for shares of a Canadian entity, the transaction will often be structured as a plan of arrangement. This is because there is an exemption from the registration requirements under US securities laws for securities issued pursuant to a transaction that a court has determined is fair to shareholders.

Disadvantages to this structure include the fact that the transaction can take longer than a takeover bid – even the negotiation of the arrangement agreement can take significant time – and can be driven largely by the target. In addition, the involvement of the courts in the process and the availability of dissent rights can result in objections to the transaction being raised.

Mergers/amalgamations: A statutory merger or amalgamation occurs where two companies combine, subject to receipt of shareholder approval, to form a new single corporate entity. The purchaser and vendor negotiate the terms of an amalgamation agreement. Proxy materials describing the nature of the transaction are then sent to the security holders for approval. Under Canadian corporate law (eg, the Canada Business Corporations Act), a merger/amalgamation generally requires approval by $66^{2}/_{3}\%$ of the target shareholders; and most Canadian corporate statutes contain provisions setting out amalgamation procedures.

Advantages to this type of transaction are that they provide increased flexibility as to structure and involve only one step to acquire 100% of the shares of the target, subject to receipt of shareholder approvals. In addition, a merger/amalgamation is not within the jurisdiction of the courts. This may be the simplest way to acquire 100% of the shares of the target.

The disadvantages are that the target and its board are involved at an early stage, these transactions usually take longer than a takeover bid and may take significant time to negotiate, they require shareholder approval and the process is driven largely by the target.

6. Conclusion

A number of factors come into play into determining the transaction structure for the purchase and sale of mineral projects – in particular, whether the transaction will be structured as an asset purchase or a share purchase. As each transaction presents its own unique facts and each party to the transaction will likely have its own set of priorities, no one factor is necessarily determinative. However, as indicated above, factors such as the stage of the project, tax planning and liabilities may help to inform the ultimate structure that the transaction takes; so too might the relative bargaining power of the parties. Once the parties have decided whether the transaction will be structured as an asset purchase or share purchase, this is not the end of the story. The parties must also navigate the various types of asset and share purchases to ultimately settle upon the appropriate structure for the particular transaction. As discussed in this chapter, there are many forms that an asset or share purchase may take, each with its own set of advantages and disadvantages.

This chapter was prepared as of 16 July 2018 and is based upon laws in effect as of that date.

Mine decommissioning and reclamation

Scot Anderson
Hogan Lovells

1. Introduction

Every mining project begins and ends with optimism. At the launch of a mining project, the mining company has identified a valuable reserve, put together a mine plan and conducted the requisite social and environmental studies. The company invests its capital in the ground, confident that the mine will provide a return on that investment.

At the end of mine life, the mine workings have been reclaimed, topsoil has been distributed across the site and the area has been reseeded. The mine area is ready for a new future and a new use.

In designing an approach to decommissioning and reclamation, the key questions are as follows:

- How can decommissioning and reclamation be incorporated into mine design and permitting?
- How can mine operations be conducted in a manner to prepare for decommissioning and reclamation?
- How can a regulator assure that a mining operation will complete decommissioning and reclamation at the end of mine life?
- What test can be used to measure successful reclamation?
- At what point can a mine operator be relieved of any further responsibility for a reclaimed mine site?

2. The lifecycle of a mine

A mining project is capital intensive, with significant capital invested before production begins. The mining company must conduct exploration to confirm and delineate its reserve, and then conduct a feasibility study to confirm that the resource can be developed economically and in compliance with regulatory requirements. Once the financial foundation has been laid for the mining project, the mining company will face another substantial investment in mine and infrastructure development before production begins. At that point, the mining company will begin to see a return on its investment. Eventually, the reserve will be fully extracted or become uneconomic. Once economic reserves have been extracted, the mining company must then decommission the mine and reclaim the mine site.[1] "A successfully completed closure process is one in which the land affected has been rehabilitated to a sustainable condition. Ideally, in this completed state, the mining lease ownership has been relinquished by the mining company to another land user with minimized continued liability."[2] The ideal outcome of mine closure and reclamation is to allow the mining company to walk away from the mine site with no further liability or responsibility. To achieve this ideal outcome, the mining company will need to do a fair amount of work on the land and show that the reclaimed site is stable and self-sustaining.

Under most mine permitting requirements, the mining company must plan for mine decommissioning and reclamation as part of the mine planning process. Even where the regulatory agency or government does not specifically require a reclamation plan as part of the initial permitting process, the decommissioning and reclamation of a mine is factored into mine planning. The costs of reclamation can be substantial and need to be part of the analysis of the financial viability of a mine. While it is possible simply to pay for reclamation out of pocket when the time comes – if the mining company has a robust balance sheet – in most instances the mining company will want to account for the costs of reclamation as part of the ongoing mining venture. Those costs should be internal to the project in most instances.

Because reclamation costs will be in the distant future and incurred only after the reserve has been extracted and monetised, the costs of reclamation might be overlooked, or at least downplayed, in the investment decision. To the extent that the mine is partially reclaimed as mining progresses, those costs get factored in alongside other capital and operating expenses. In most cases, however, much of the cost of decommissioning and reclamation will be incurred 10, 20 or 50 years in the future. When the mining company considers the net present value of those costs, they become diminishingly small, even if the real dollar costs in the future are substantial. Also, some investors in a mining project may plan to exit the project well before the end of mine life, which means those investors will not be around when the reclamation bill comes due. The remoteness of mine closure, decommissioning and reclamation can lead mining companies and investors to put off thinking about those costs or fail to have a plan to pay for them. If the project has failed to plan for closure, the regulators may be faced with a site with substantial reclamation costs and an insolvent mining company. Much of the regulatory approach to reclamation described in this chapter is designed to avoid precisely that outcome

The Fraser Institute provides a useful summary of the steps in mine closure:

---

1 Evelyn L Jessup Bingham, "Closure Planning," *SME Mining Engineering Handbook*, Ch 16.7 at 1753 (vol 2, 2011).

2 *Id* at 1754.

- Shut-down: Once production stops, the number of workers is reduced and only a small labour force is retained to permanently shut down the mining equipment.
- Decommissioning: Small crews or contractors decommission or take apart the mining processing facilities and equipment. Pipelines are drained, equipment and parts are cleaned and sold, buildings are repurposed or demolished, warehouse materials are recovered and waste is disposed of.
- Remediation/reclamation: The objective of reclamation is to return the land and watercourses to an acceptable standard of productive use, ensuring that any landforms and structures are stable and any watercourses are of acceptable water quality.
- Post-closure: Monitoring programmes are used to assess the effectiveness of the reclamation measures and to identify any corrective action that may be needed. In addition, mines may require long-term care and maintenance after mine closure, such as ongoing treatment of mine discharge water, periodic monitoring and maintenance of tailings containment structures, and monitoring any ongoing remediation technologies used such as constructed wetlands.[3]

This outline shows the progression of closure and reclamation from a functioning mine to reclaimed lands. The outline also captures an important concept in how to think about reclamation. The goal of reclamation is to "return the land and watercourses to an acceptable standard of productive use". Reclamation is not restoration – the goal of reclamation is not to restore lands to their prior status and productivity, but rather to reclaim lands to a point where they function in a manner consistent with their intended use. If a reclaimed area will be used for an industrial facility after mining ceases, then the measure of reclamation success will be consistent with that use and will require less revegetation (for example) than a site that will be used for agriculture or wildlife habitat. To be clear, there are many circumstances where reclaimed lands have soil stability and vegetative cover that is better than nearby lands. Because the miner will pick the seed mix for revegetation and will designed the configuration of backfill and soil distribution, the reclaimed lands may be more productive than native areas. Still, the important point is that the goal of reclamation is to create lands suitable for their intended purpose and not to restore those lands to their original condition.

3 www.miningfacts.org/what-happens-to-mine-sites-after-a -mine-is-closed.

3. Basic decommissioning and reclamation standards

Modern mine closure standards were first adopted in the 1970s in the United States, Canada, Australia and some European countries.[4] The standards and strategies adopted in the 1970s have evolved over the years. Most countries adopting comprehensive reclamation standards have used the key concepts developed in the 1970s. Still, a number of countries with mining operations have relatively simply guidelines for closure and reclamation of mining sites. As Luke Danielson notes:

> *The mining industry has globalized, but this model of closure legislation has not. Administering such legislation well requires a degree of capacity not always present in governments, a degree of public pressure that is not always there, and money that is not always available. So while there are increasing closure requirements in many countries, they have not taken on all of the characteristics of the fully developed systems. Most strikingly, many of the closure systems in these countries do not have a system of financial guarantees, so when industry prices fall and companies fail (which happens fairly often in this risky industry), there may be no funds to implement closure. Companies in jurisdictions that require rigorous closure planning and financial assurance are in direct economic competition with operations in places that have few or even no requirements.*[5]

A mining company assessing where to invest its capital faces something of a dilemma when looking at fully developed reclamation regimes as compared to underdeveloped reclamation regimes. There is always a risk of a race to the bottom – that is, there is a temptation to select an investment in a country with relatively relaxed reclamation requirements, because that approach can reduce reclamation costs and provide flexibility. This is Danielson's point about competition. The lack of a regulatory regime has only a superficial appeal, however. The lack of clearly defined standards creates uncertainty and risk. When there is no legal or regulatory measure of success, the regulator or government can act on a whim. At the end of mine life, the regulator may face pressure from local communities to step in and impose unexpected standards or requirements on the mine operator. It is important to have a legal framework for closure and reclamation. Without pre-existing legal standards, almost any standard can be imposed. Thus, there is great utility in selecting a site for a mining project with a robust and well-established closure and reclamation regime. With knowledge of legal requirements at the beginning of the mining venture, the mining company and investors can quantify and plan for the costs and timing of closure and reclamation. That certainty is very valuable.

---

4 Luke Danielson and Marketa Zubkova, "Integrated Mine Closure Planning in the Framework of Sustainable Development", Rocky Mt Min L Fdn, Special Institute: *Mine Closure, Financial Assurance and Final Reclamation*, Ch 10 (2009).

5 *Id.*

4. Coal

The United States adopted a comprehensive approach to the decommissioning and reclamation of coal mines with the passage of the Surface Mining Control and Reclamation Act in 1997.[6] The Surface Mining Control and Reclamation Act controls the potential environmental effects of coal mining through a comprehensive regulatory scheme based on permitting prior to the commencement of mining, performance standard applicable during mine operations and requirements for reclamation.[7]

First, a coal company is required to include a reclamation plan as part of its permit application.[8] The reclamation plan must provide "the degree of detail necessary to demonstrate that reclamation...can be accomplished".[9] The reclamation plan must specify, for example:

- the lands to be mined and the size and timing of mining operations;
- the condition of the land prior to the mining application, including uses prior to mining, and the capability and productivity of the land;
- the use to be made of the land after reclamation;
- how the post-mining land use will be achieved;
- the engineering techniques to be used in mining and reclamation, including plans for control of surface water drainage, backfilling, soil stabilisation and revegetation;
- consideration of maximum recovery of the coal resources;
- a detailed timetable for the accomplishment of each major step in the reclamation plan;
- consideration of surface owner plans and state and local land use plans;
- steps to comply with applicable air and water quality laws and health and safety standards;
- consideration of the local physical environment and climatological conditions; and
- a detailed description of measures taken during mine operations and reclamation to assure the protection of surface and ground water, including water users and water quality.[10]

One of the key provisions of the Surface Mining Control and Reclamation Act is the requirement that the mining company post a performance bond to assure that the coal mine will be reclaimed. After the permit is issued, but prior

6 In the United States, hard rock mining remains governed by the General Mining Law of 1872, and thus reclamation tends to be managed under state statutes.

7 Joseph J Zaluski and Lesly AR Davis, "Introduction to the Surface Mining Control and Reclamation Act", 16 E Min L Found § 8.01 (June 1995).

8 Surface Mining Control and Reclamation Act, 30 USC § 1257(d).

9 *Id*, 30 USC § 1258(a).

10 *Id*, 30 USC § 1258.

to commencing mining operations, the mining company must post a performance bond.[11]

The bond will be sufficient to cover the cost of reclamation if the mining company fails to complete the reclamation work. The bond is adjusted during the life of the mine to reflect the changing cost of reclamation.[12] The bond is based on the reclamation plan submitted with the permit and the probable difficulty of reclamation in light of such factors as topography, geology, hydrology and revegetation potential.[13]

The bond is often in the form of a surety bond – that is, a promise from a surety company to fund the amount of the bond should the regulator determine that the mining company will not complete reclamation.[14] The regulatory agency will also accept a collateral bond – a bond secured, for example, by cash, real property or a letter of credit.[15] Finally, the regulator may also accept a self-bond. A mining company may be allowed to provide security for reclamation based on its balance sheet. If a company has sufficient financial resources and is a good credit risk, the regulator may simply accept a promise from the company that it will perform reclamation when the time comes.[16]

Historically, most US coal mining companies used surety bonds to fulfil the reclamation bonding requirements of the Surface Mining Control and Reclamation Act. At the turn of the century, however, the surety bond market experienced financial losses, resulting in consolidation in the surety industry and the dissolution of several surety companies.[17] As a result, surety companies were less willing to provide bonds for mining reclamation and often required cash collateral for a surety bond.[18] Providing cash collateral can create a significant negative effect on the economics of a mining project.

The Surface Mining Control and Reclamation Act also applies performance standards during mine operations to promote final reclamation. These performance standards include the following:

- Topsoil is segregated from other materials and placed in a separate marked pile. The pile is seeded with quick-growing plants to prevent erosion of the topsoil stockpile.[19]
- Reclamation is to occur "as contemporaneously as practical".[20] This requirement means that the mining company will conduct reclamation as it goes along, rather than waiting until the end of the mining to start

11 *Id*, 30 USC § 1259(a).
12 *Id*, 30 USC § 1259(a),(e).
13 *Id*.
14 *Id*, 30 USC § 1259(a).
15 *Id*.
16 *Id*, 30 USC § 1259(c).
17 See Donald Vagstad and Laurence J Kirsch, "Addressing the Crisis in Reclamation Bonding for Mining Companies," 48 Rocky Mt Min L Inst Ch 17 (2002).
18 *Id*.
19 30 USC § 1265(b)(5).
20 30 USC § 1265(b)(16).

reclamation. For a surface mine, for example, the mining company will backfill the pit behind it as mining advances. For an underground mine, it means that spoil piles, abandoned roadways and other surface uses will be reclaimed when it is practicable to do so.

- Mining operations must be conducted in a manner that assures mine stability and prevents slides or waste accumulation.[21] Doing that assures that mine disturbances are minimised, thereby reducing the amount of land requiring reclamation.

When it is time to complete reclamation, the Surface Mining Control and Reclamation Act provides fairly detailed requirements for the reclamation process:

- The most important requirement is that reclamation be conducted to support the post-mining land use.[22] As noted above, the mining permit will specify the land use for the mine area once mining is complete. The area might be used, for example, for wildlife habitat, agriculture, residential or industrial uses. The reclamation plan will use a seed mix, revegetation plan, topography and design consistent with the specified past-mining land use. The specified use can be the prior use of the area or "higher or better uses".[23]
- One of the more controversial provisions of the Surface Mining Control and Reclamation Act is the requirement that the reclaimed area be graded "in order to restore the approximate original contour of the land with all highwalls, spoil piles, and depressions eliminated".[24] Achieving this goal can be difficult and expensive. For a surface mine, the backfill placed in the mine pit expands and the mining company may have a volume in excess of the void to be filled. In other circumstances, the overburden used to backfill a mine may be thin, leaving an insufficient volume of material to allow backfilling to the approximate original contour.

There is a specific exception from the approximate original contour requirement in the Surface Mining Control and Reclamation Act for mountain top removal, which creates flat areas in mountainous terrain.[25]

- Final reclamation must protect the prevailing hydrologic balance at the mine site and associated offsite areas.[26] This requirement applies to surface and groundwater.

---

21 30 USC § 1265(b)(21).
22 30 USC § 1265(b)(2).
23 *Id.*
24 30 USC § 1265(b)(3).
25 30 USC § 1265(c).
26 30 USC § 1265(b)(10).

- Any waste piles must be compacted, stabilised and revegetated.[27]
- All disturbed areas will be revegetated with "a diverse, effective and permanent vegetative cover of the same seasonal variety nature to the area . . . at least equal in extent to cover to the natural vegetation in the area".[28]
- For underground mines, all portals and entryways must be sealed.[29]

A mine operator remains subject to the jurisdiction of the regulator until reclamation is successfully completed. The mine operator must maintain its reclamation bond until reclamation is finished, although the bond can be partially released as key reclamation requirements such as backfilling are completed. The Surface Mining Control and Reclamation Act contemplates the release of 60% of the mining company's bond upon completion of backfilling, regarding and drainage control.[30]

A mining company can secure complete bond release when the mine area no longer contributes suspended solids to stream flow or runoff, and when any affected prime farmland returns to its prior productivity.[31] Also, it is not enough simply to have reseeded the affected area. Rather, the mining company must show vegetative success for five years in areas with more than 26 inches of average annual precipitation and 10 years for more arid areas.[32] After establishing successful revegetation, the mining company can have its entire bond released. Of equal significance, the mining company is no longer subject to the jurisdiction of the mine regulator. The mining company is relieved of any further responsibility for mine reclamation.[33]

While the Surface Mining Control and Reclamation Act applies only to coal mines in the United States, this regulatory regime establishes many of the key concepts affecting the decommissioning of mines and the reclamation of mined lands globally:

- The mine permitting process should include a reclamation plan.
- The mine should be designed in a way to promote successful reclamation.
- The mining company will provide a financial guaranty to assure that the reclamation will completed even if the mining company refuses to conduct reclamation.
- The reclamation plan will be designed to promote and facilitate a selected post-mining land use.

---

27 30 USC § 1265(b)(11).
28 30 USC § 1265(b)(19).
29 30 USC § 1266(b)(2).
30 30 USC § 1265(c)(1).
31 *Id.*
32 30 CFR § 816.116.
33 See, for example, *Navajo Nation*, IBLA 2000-241 (26 Oct 2004) (confirming that regulatory agency can terminate its jurisdiction over lands where reclamation is successfully completed).

- The reclamation plan will be designed to protect the prevailing hydrologic balance.
- The mine design, including the reclamation plan, will be designed to promote maximum recovery of the mineral reserve.
- Reclamation should occur as contemporaneously as practicable with ongoing mining operations.
- The reclaimed area will be revegetated with plants similar to native vegetation, assuring vegetative cover equal to pre-existing cover.
- Topsoil will be segregated and used to promote reclamation and revegetation.
- Stockpiles, tailing piles and slopes and highwalls should be reclaimed to assure stability.
- Backfilling and grading should create a topography consistent with post-mining land use.
- The regulator should retain jurisdiction over the mine operator until reclamation is successfully completed.

5. Hard rock mining

The United States does not have a comprehensive federal statute requiring the permitting and reclamation of hard rock mines. As a result, the decommissioning of non-coal mines is largely left to the states.[34] State laws on hard rock mine permitting and reclamation often follow the conceptual framework created by the Surface Mining Control and Reclamation Act.

South Dakota and Arizona are two states with a long history of hard rock mining. Both states have mine reclamation acts that track the Surface Mining Control and Reclamation Act model and apply it to hard rock mining. Both Arizona and South Dakota require a permit to conduct hard rock mining operations, and the permit application must include a reclamation plan.[35]

In South Dakota, the reclamation plan must specify the post-mining land use, with that use developed by the mining company, the landowner and the regulatory agency.[36] Similarly, in Arizona, the reclamation plan must specify the proposed post-mining use of the land.[37] Each statute includes requirements for the reclamation plan to address topography,[38] water and hydrologic balance,[39] wildlife protection,[40] revegetation[41] and spoil pile stability.[42]

---

34 American Law of Mining § 173.03(i) (2d ed 2017).
35 See Ariz Rev Stat Ann § 27-921; SD Codified Laws § 45-6B-5.
36 SD Codified Laws § 45-6B-44.
37 Ariz Rev Stat Ann § 27-971(B)(4).
38 Ariz Rev Stat Ann § 27-971(B)(6); SD Codified Laws § 45-6B-37 (topography to be consistent with post-mining land use).
39 Ariz Rev Stat Ann § 29-973(B)(2) (reclamation plan to protect water resources and water management projects); SD Codified Laws § 45-6B-41 (minimise disturbance of hydrologic balance).
40 Ariz Rev Stat Ann 971(b)(9)(d); 973(B)(3); SD Codified Laws §§ 45-6B-44, -45, -92.
41 Ariz Rev Stat Ann 27-971(b)(9)(c), -974; SD Codified Laws § 45-6B-39.
42 Ariz Rev Stat Ann §§ 27-901(13); 27-971(9)(b); SD Codified Laws § 45-6B-43.

In addition, both Arizona and South Dakota require a hard rock mine operator to post a bond to assure final reclamation of a mine. Arizona provides considerable flexibility as to the form of financial assurance. Arizona will accept any combination of the following:

- surety bond;
- certificate of deposit;
- trust fund with pay-in period;
- letter of credit;
- insurance policy;
- certificate of self-insurance;
- cash deposit with the state treasurer;
- evidence of ability to meet a corporate financial test or corporate guarantees as provided by 40 Code of Federal Regulations Section 264.143(f);
- annuities; and
- additional financial assurance mechanisms that are acceptable to the inspector.[43]

South Dakota has a preference for a surety bond, but will consider other factors such as cash collateral or a company's net worth.[44]

The surety bond will be sufficient to allow reclamation of the mine operation and may be adjusted during the life of the mine to reflect changes in the amount of reclamation required at any given time during mine operations.[45]

Finally, for both states, the bond may be released only after the mine operator demonstrates successful reclamation. In Arizona, the bond can be partially released where surface disturbances have been reclaimed, but final bond release cannot occur until revegetation success has been demonstrated for up to three growing seasons.[46] Similarly, in South Dakota, the regulator will release the surety bond only upon completion of reclamation.[47] South Dakota may find that a mine requires post-closure care and may require an operator to keep some portion of the bond in place for up to 30 years.[48]

The reclamation laws of Arizona and South Dakota are offered as examples of how the reclamation framework established under the Surface Mining Control and Reclamation Act has been translated to hard rock mining. Other states have adopted similar laws.[49]

---

43 Ariz Rev Stat Ann § 27-991.
44 SD Codified Laws §§ 45-6B-20.1 to -23.
45 Ariz Rev Stat Ann § 27-995; SD Codified Laws § 45-6B-21.
46 Ariz Rev Stat Ann § 27-996(B).
47 SD Codified Laws § 45-6B-25.
48 SD Codified Laws § 45-6B-91.
49 See, for example, American Law of Mining § 173.03 (describing hard rock reclamation laws in Colorado and Nevada).

6. Uranium mining

Uranium mining and the mining of radioactive minerals can present special issues for reclamation:

> *Uranium ores and mill tailings are a particular health concern because they emit radionuclides, especially radon-222 which is emitted as a gas and can be released into the atmosphere when the uranium-bearing materials are brought to the surface. The decay products of radon-222 can attach to dust particles and increase the risk of lung cancer when inhaled. Mill tailings also release radon-222, unless covered, and once dry and exposed can provide a source of windblown radioactive dust particles.*[50]

As a result, the reclamation of uranium mining sites requires addition steps to manage radioactivity. Even though the uranium has been removed from uranium mill tailings, those tailing continue to emit radioactivity in the form of radon-222 and the dust from the tailings can spread radioactive materials.

In the United States, the Environmental Protection Agency (EPA) requires radium concentrations in soil to be reduced to levels of no more than 5 average picocuries per gram in the first 15 centimetres (cm) of soil and in each subsequent 15 cm layer, averaged over 100 square metre segments. The EPA also requires that disposal systems be designed to provide "reasonable assurance" of achieving the radon emission standard for 1,000 years without the need for active maintenance.[51] Tailing sites may also require special measures to protect groundwater, such as lined impoundments.[52]

7. Reclamation policies from non-governmental organisations

In addition to regulatory regimes developed by governments, several trade groups, non-governmental organisations and international lending agencies have put forward policies and guidance related to mine closure and reclamation. For example, the International Council on Mining and Metals (ICMM), in its Sustainable Development Framework: ICMM Principles,[53] seeks to promote sustainable development in mining operations. One of its key principles is to "[i]ntegrate sustainable development in corporate strategy and decision making processes", and more specifically to assure that a mining company plans, designs and operates and closes a mine in a manner that enhances sustainable development.[54] The ICMM also recommends an assessment of environmental impacts "from exploration to closure", and to

---

50 Gregory P Williams, "Closing a Hardrock Mine—Can You Ever Walk Away?", 39 Rocky Mt Min L Inst Ch 10 (1993)

51 Anthony J Thompson and Christopher S Pugsley, "United States Federal/State Licensing/Permitting Regimes Implicated By Uranium Development", Rocky Mt Min L Fdn Special Institute: *Uranium Exploration and Development*, Paper 3 (2006).

52 *Id.*

53 Sustainable Development Framework: ICMM Principle (rev ed 2015). The ICMM is a consortium of the major international mining companies.

54 *Id* at 3.

have "adequate resources" to assure closure.[55] While the ICMM principles are fairly broad and high level, the principles are significant because they show a commitment from major mining companies to implement many of the key reclamation concepts, even in jurisdictions lacking a comprehensive reclamation regime.

The ICMM principles also note the importance of collaborating with host communities.[56] That concept has been further developed by the Mining Association of Canada in its Mine Closure Framework developed as part of its "Toward Sustainable Mining" initiative.[57] The Mine Closure Framework recommends working with communities of interest to develop a closure plan and strategies to mitigate impacts and provide for long-term economic development.[58] The mining company is to work with the communities of interest to identify values important to the community and incorporate those values into the reclamation plan.[59] The Mine Closure Framework also adopts many of the other key reclamation concepts.[60]

Finally, a mining project may be subject to the Equator Principles if the mining company secures financing from one of 92 financial institutions that have adopted the Equator Principles.[61] While the Equator Principles are voluntary, in the sense that there is no law imposing them, they are effectively mandatory for mining companies seeking financing from an Equator Principles Financial Institution (EPFI). The Equator Principles require projects with potential adverse social or environmental impacts (Category A or B projects) to conduct a social and environmental assessment as part of mine financing, and to prepare reports on those impacts over time. The Equator Principles require a specific covenant in financing documentation "to decommission the facilities, where applicable and appropriate, in accordance with an agreed decommissioning plan".[62]

In addition to the Equator Principles, the International Finance Corporation (IFC) (part of the World Bank Group) has developed Performance Standards on Environmental and Social Sustainability.[63] While, like the Equator Principles, the IFC Performance Standards do not have the force of law, host countries often include a general requirement to comply with the Performance Standards as part of a concession agreement or licence. Also, the Performance Standards may be incorporated in financing agreements. As with the Equator Principles,

55 *Id* at 5.
56 *Id* at 6.
57 The Mining Association of Canada, *Towards Sustainable Mining Mine Closure Framework* (Nov 2008).
58 *Id*, §3.
59 *Id*, §2.
60 *Id*, §2, 48.
61 www.equator-principles.com/members-reporting
62 Equator Principles, Principle 8(d).
63 International Finance Corporation, *Performance Standards on Environmental and Social Sustainability* (1 January 2012).

the IFC Performance Standards require preparation of an environmental and social impact assessment.[64] This assessment may include an assessment through the entire lifecycle of a project, from design to closure and post-closure.[65] Decommissioning is to be conducted in accordance with good international industry practice.[66]

8. Acid rock drainage

The reclamation of hard rock mines must also take into consideration the management and mitigation of acid rock drainage (ARD). The mining process may create surface piles of tailings and overburden, as well as open adits and other underground voids. When this waste rock is exposed to water, the water picks up sulphuric acid from the weathering of iron pyrites and other sulphuric minerals. The acidic water will then mobilise and transport heavy metals.[67] A mine operator will design a mine to minimise the production of ARD during mine operations.[68] In the reclamation process, the mine operator can also take steps to prevent or minimise the generation of ARD. These approaches include the following:

- *Chemical treatment of tailings: tailings can be treated before they are discharged to waste repositories. For example, acid-leached tailings can be neutralized to raise the pH and lower metals solubility*

  . . .

  Selective waster placement: waste repositories should be designed so that transport pathways that may allow contaminants to migrate out of the water are minimized

  . . .

- *Up-gradient water diversion: in locations where surface or groundwater comes into contract with chemically reactive waste materials or wall rock, it is beneficial to capture the clean up-gradient water before it contacts the reactive material. This will reduce the likelihood of contaminants being mobilized and will protect water quality from being degraded*

  . . .

- *Selective handling/encapsulation: chemically reactive mineral waste may be selectively handled and surrounded with benign waste to limit the contact of runoff water, infiltration, and groundwater*

  . . .

- *Blending/mixing: the blending and mixing of acid-generating wastes with net*

---

64 *Id*, Performance Standard 1.

65 *Id* at § 14.

66 *Id* at § 6.

67 See Hadassah Reimer and Murray Feldman, "To Fill, or Not to Fill: Reclamation and Closure Considerations for Open Pit Mines", 58 Rocky Mt Min L Inst 15 (2012).

68 See Rens Verberg, "Mitigating Acid Rock Drainage," *SME Mining Engineering Handbook*, Ch 16.5 (3d ed 2011).

neutralizing materials may be used to ensure that bulk neutral pH conditions are maintained within the waste mass as a whole, either in perpetuity or for a sufficiently long period to allow another long-term management technique to be implemented

. . .

- *Liners: liners can be constructed under reactive mineral wastes to allow capture of contact water that would normally discharge to the surface or to groundwater*

. . .

- *Micro-encapsulation (passivation): this process involves the coating of mine wastes or pit walls with products that encapsulate sulfide surfaces and inhibit oxidation reactions.*[69]

Many of these strategies can be implemented during mine operations and doing so will make final reclamation easier to achieve. Still, the presence of ARD or other reactive chemicals can be a serious impediment to final reclamation. Post-closure water management is commonly required for disposal facilities that contain chemically reactive mineral waste. Contaminated water that discharges from the waste material will typically need to be collected and treated before it can be discharged to the environment. Depending on the rates at which sulphide minerals oxidise and at which soluble constituents migrate out of the waste, water management may be required for decades or even centuries after closure. Spent heap leach pads containing hazardous process fluids may need to be flushed with many pore volumes of clean water to meet water quality discharge requirements. Long-term water management costs can be high, in many cases exceeding the cost of the initial rehabilitation. Post-closure monitoring programmes are also usually required to ensure that surfaces are not eroding, the mandated vegetation is permanently established, and chemical exposure risks and impacts remain within acceptable ranges. If only inert waste is present, post-closure monitoring may be completed within a few years of closure; but if chemically reactive waste is present, monitoring may be required for many decades.[70] Mining companies will incorporate passive treatment systems and similar strategies to avoid long-term treatment of ARD.[71]

9. Artisanal mining

Artisanal and Small-scale Gold Mining (ASGM) involves mineral extraction using manual labor and rudimentary tools. It requires low investment and little

---

69 Richard K Borden, "Waste Disposal and Contamination Management", *SME Mining Engineering Handbook*, Ch 16.6 (Table 16.6-1, Common Management Strategies for Chemically Reactive Mineral Waste).

70 *Id* at 1745.

71 Dalva L Moellenberg, "Technical, Legal, and Financial Strategies for Effective Mine Closure", 51 Rocky Mt Min L Inst Ch 18 § 18.03[3] (2005).

infrastructure. It may be a full time occupation, seasonal, or may provide a supplementary income. It can be a coping mechanism when other livelihoods fail or a preferential livelihood which surpasses other opportunities. Activities may fluctuate with commodity prices with high prices making mining more rewarding. Relative to developed world incomes, and incomes of the middle class in developing countries, incomes in ASGM are moderate to low for most individuals, but for those individuals, incomes are relatively high compared to alternatives. Frequently five times the value of alternatives.[72]

Artisanal or illegal mining can be an impediment to reclamation. These activities often take place near or within the formal mining concessions of large-scale mining operations. Artisanal mining may take place in abandoned mining areas, in tailing dams or in downstream areas.[73]

Thus, the existence of artisanal mining at or near a larger mine can create ongoing operational impediments, including the discharge and release of chemicals that the legal mine cannot control. Also, if artisanal miners are mining in tailings areas or areas where mining has been completed, the mining company will face resistance to the reclamation of those areas of the mine.

The mining company may have no choice but to find a way to incorporate and adopt ongoing artisanal mining into mining and reclamation plans. The government of Ecuador, for example, recognised that artisanal mining is an important source of income for certain communities and thus passed a mining law designed to bring artisanal mining under at least a modicum of regulation and oversight.[74] Similarly, the Africa Mining Legislative Atlas provides several examples from Africa of mining legislation designed to bring artisanal mining into the realm of regulation and thereby provide better environmental oversight.[75]

10. Key reclamation concepts in representative jurisdictions

The development of comprehensive closure and reclamation laws in the 1970s, coupled with the advancement of those laws over the intervening time, has established the key concepts foundational to an effective regime for mined land reclamation. For ease of reference, those key concepts are as follows:

- The mine permitting process should include a reclamation plan.
- The mine should be designed in such a way as to promote successful reclamation.

---

72 Kevin Telmer and Anthony Persaud, "Historical and Modern Government Responses to Artisinal and Small Scale Gold Mining", Rocky Mt Min L Fdn, Special Institute: *International Mining and Oil & Gas Law, Development, and Investment*, Ch 18C (Apr 2013).

73 www.miningfacts.org/communities/what-is-artisanal-and-small-scale-mining/.

74 See generally Andrew A Irvine, "Artisanal and Small-Scale Mining in Ecuador—Building and Implementing an Effective Legal Framework", Rocky Mt Min L Fdn, Special Institute: *International Mining and Oil & Gas Law, Development, and Investment*, Ch 20A (Apr 2017).

75 See Africa Mining Legislative Atlas, Part B-4, www.a-mla.org.

- The mining company will provide a financial guaranty to assure the reclamation will completed even if the mining company refuses to conduct reclamation.
- The reclamation plan will be designed to promote and facilitate a selected post-mining land use.
- The reclamation plan will be designed to protect the prevailing hydrologic balance.
- The mine design, including the reclamation plan, will be designed to promote maximum recovery of the mineral reserve.
- Reclamation should occur as contemporaneously as practicable with ongoing mining operations.
- The reclaimed area will be revegetated with plants similar to native vegetation, assuring vegetative cover equal to pre-existing cover.
- Topsoil will be segregated and used to promote reclamation and revegetation.
- Stockpiles, tailing piles and slopes and highwalls should be reclaimed to assure stability.
- Backfilling and grading should create a topography consistent with post-mining land use.
- The regulator should retain jurisdiction over the mine operator until reclamation is successfully completed reclamation laws that serve as a benchmark for best practices in mining reclamation.

The government of Australia adopts a definition of 'mine reclamation' that is generally consistent with the best practices:

Rehabilitation comprises the design and construction of landforms as well as the establishment of sustainable ecosystems or alternative vegetation, depending upon desired post-operational land use.

Mine site rehabilitation should be designed to meet three key objectives:

1. *the long-term stability and sustainability of the landforms, soils and hydrology of the site*
2. *the partial or full repair of ecosystem capacity to provide habitats for biota and services for people*
3. *the prevention of pollution of the surrounding environment.*[76]

It is useful to look at a few important mining jurisdictions to see how mining reclamation regulations compare to these key principles. Many mining jurisdictions have not had regulations or laws specifically addressing mine closure. As these laws come into place, they tend to adopt the concepts set out as the key principles. There is nonetheless a fair amount of variability among

76 Australian government, "Mine Rehabilitation: Leading Practice Sustainable Development Program for the Mining Industry" at 2.0 (September 2016).

jurisdictions concerning the reclamation requirements, and some jurisdictions have not adopted reclamation or closure laws of the scope or scale necessary to assure successful reclamation of mine sites.

10.1 Australia

Mine closure and reclamation is managed at the state level in Australia, but the Australian government sets forth an approach to mine reclamation that is consistent with the key principles of reclamation law. Reclamation or rehabilitation is addressed early in the mining process and managed through mine closure.[77] Under the process described by the Australian government, the mining company and regulators will set rehabilitation targets and objects at the beginning of mine planning. These targets and objectives will be used for rehabilitation planning through the permitting and design process. During the course of mine operation and as part of the closure process, the mining company will use rehabilitation techniques to assure effective closure:

- landform design and construction;
- reconstruction of the soil profile;
- selection of suitable species for mine rehabilitation;
- establishment of vegetation; and
- fauna recolonisation.

As the mining company works through these steps in the rehabilitation process, the mining company and regulators will measure reclamation success against a specified set of rehabilitation criteria, and will manage and monitor reclamation success against those criteria. The criteria for success include both qualitative and quantitative measures designed to assure that reclamation is complete and sustainable.[78]

Australian states also use surety bonding and financial guarantees to assure completion of reclamation.[79]

10.2 Peru

Peru passed a Mine Closure Act in 2003 and followed that legislation with a comprehensive set of regulations in 2005.[80] The Mine Closure Act and the related regulations require a mine closure plan as part of the mine permitting process, as opposed to the prior legislation, which did not require a closure plan

77 Australian government, "Mine Rehabilitation: Leading Practice Sustainable Development Program for the Mining Industry" at 3.0 (September 2016).

78 *Id* at 3.2.

79 James Otto, "Global Trends in Mine Reclamation and Closure Regulation", Rocky Mt Min L Fdn, Special Institute: *International Mining and Oil and Gas Law, Development, and Investment*, Ch 9A (2007); Mineral Council of Australia, "Mine Rehabilitation in the Australian Minerals Industry" (2015).

80 Juan Carlos Escudero Velando and Fernando Valencia Dibos, "The Main Legal and Financial Aspects related to Mine Closure in Peru", Rocky Mt Min L Fdn, Special Institute: *International Mining and Oil & Gas Law, Development, and Investment*, Ch 9B (Apr 2007).

until after mining was completed.[81] This change makes successful reclamation much more likely, because the mine is always working toward its reclamation goals, including reclamation contemporaneously with mining operations. The Peruvian regulations exempt small mines and exploration activities from the reclamation planning requirements by stating that the requirement of a closure plan applies only to mines that remove over 10,000 tonnes of material or over 1,000 tonnes of acidic material.[82]

Peru was also the first Latin American country to impose financial assurance requirements for mine reclamation.[83] The mine operator must prepare a budget for reclamation, including all direct and indirect costs.[84] The inclusion of indirect costs is an important and sophisticated addition to the regulations. It can be easy to underestimate reclamation cost if the focus is solely on materials or labour. Oversight and management of the process are also crucial to its success. This budget is adjusted three years after the approval of the closure plan and then every five years thereafter.[85] This amount then serves as the basis of the reclamation bond. Consistent with the best practices described above, the bond must be in a form where the cash required for reclamation is easily accessible (and liquid), and sufficient to complete reclamation in the event of a default by the mine operator.[86] The Closure Act is flexible and allows security to be posted in the form of cash, trust accounts, mortgages over property and similar surety instruments. The regulations, however, restrict the forms of surety to surety bonds, letters of guarantee from a bank and similar third-party financial instruments.[87] In 2006 Peru enacted a new law that allows a mining company to guarantee its reclamation by engaging in reforestation activities.[88]

Also consistent with the key concepts outlined above, Peru does allow partial release of reclamation bonds as reclamation progresses. Total bond release cannot occur until all the requirements of the closure plan are met.[89] The mining company must maintain a portion of the bond for at least five years post-closure, to assure that reclamation has been successful.[90]

The Peruvian law is sophisticated and well developed. It adopts many of the key concepts applicable to the regulation of closure and reclamation, and should assure that mining projects in Peru are reclaimed successfully.

---

81 *Id.*
82 *Id.*
83 *Id.*
84 *Id.*
85 *Id.*
86 *Id.*
87 *Id.*
88 *Id.*
89 *Id.*
90 *Id.*

10.3 Chile

Chile passed a comprehensive law on mine closure and reclamation in 2011, with related regulations promulgated in 2012.[91] The new Chilean law requires the development of closure plans as part of mine operations and establishes a programme for the posting of financial assurances for closure, in the form of a reclamation fund.[92] The mining regulator in Chile, SERNAGEOMIN,[93] characterises the new law as requiring that all mining sites have a closure plan approved by the SERNAGEOMIN prior to the start of mining operations. SERNAGEOMIN sees closure plan as an engineering project in which a set of measures and actions is designed to mitigate the effects of mining extractive industry, in the places where it is carried out, in order to ensure the physical and chemical stability of the facilities, in accordance with the applicable environmental regulations.[94]

The amount of financial assurance is calculated based on the cost of mine closure (on a net present value basis) and, in addition to the direct cost of closure, includes administrative and monitoring costs and a contingency factor.[95] SERNAGEOMIN does contemplate the partial release of bonded amounts, and a recalculation of the amount of financial surety, over the life of the mine.[96] The Chilean law also provides for auditing of the closure plan and the related financial assurance by public auditors.

10.4 Africa Mining Legislative Atlas

The World Bank and the Africa Legal Support Facility maintain the Africa Mining Legislative Atlas, a compendium of mining laws throughout Africa.[97] A review of the Africa Mining Legislative Atlas shows wide variability in the mining reclamation laws among African countries. The African Mining Legislative Atlas includes a Guiding Template, designed to provide examples of mining legislation that might be adopted by African countries. The Guiding Template includes provisions from the mining laws of South Africa and Uganda as examples of reclamation laws.[98] South Africa and Uganda, taken together, recommend a mine reclamation law that incorporates mine reclamation at the licensing stage, including the identification of existing land use, and a proposed future land use. Both examples include a requirement for a bond or financial assurance. The South Africa model also anticipates a closure certification process at the end of mine life.[99]

---

91 Law 20.551 (October 2011), regulations promulgated under Supreme Decree 41 (November 2012).
92 Rodrigo Munoz, *Getting the Deal Done: Mining in Chile* (June 2017).
93 *Servicio Nacional de Geología y Minería.*
94 www.sernageomin.cl/cierre-de-faenas-mineras/.
95 SERNAGEOMIN, "Methodological Guide for the calculation, determination and disposition of the financial guarantee established by Law 20551" (2018).
96 *Id.*
97 www.a-mla.org.
98 www.a-mla.org/guidingtemplate/part/292.
99 *Id.*

Many of the mining laws in Africa, however, have only very general provisions related to mine reclamation. The Botswana Mining Law of 1999, for example, requires the holder of a mineral concession to ensure that the concession area is rehabilitated "from time to time" and ultimately reclaimed "in so far as is practicable" in a manner acceptable to the director of mines. Further, the holder of a mineral concession shall make adequate ongoing financial provision for compliance with these obligations.[100] Similarly, the Zambia Mining Code simply requires "the rehabilitation, leveling, re grassing, reforesting or contouring of such part of the land over which the right or license has effect as may have been damaged or adversely affected by exploration operations, mining operations or mineral processing operations".[101] Zambia does not require financial assurances. Namibia does not have a specific provision for mine reclamation, but does have a provision stating the mine operators are responsible to remedy pollution.[102] Ethiopia requires a closure certificate, but even after closure the mine operator remains responsible for monitoring environmental effects after closure of the mine and remains liable for duties arising during the term of the licence.[103] Ethiopia requires that the licensee allocate funds to cover the costs of rehabilitation of environmental impact.[104]

11. Conclusion

The mining industry has been at the forefront of developing strategies to assure the sustainable development of mineral resources. The key document establishing the framework for this concept – *Breaking New Ground*, the Mining, Minerals and Sustainable Development report for the International Institute for Environment and Development – recognises that mine closure objectives and impacts should be considered from the project inception, with the mine closure plan defining a vision of the end results to be achieved and actions to do so, in such a manner that:

- future public health and safety are not compromised;
- the natural capital will be ensured after the mine closes;
- the use of the site will be beneficial and sustainable after mine closure, which involves land reclamation to restore or rehabilitate the land affected for further economic use;
- negative socio-economic impacts are minimised; and
- the full social and economic benefits of the project will be captured to develop the region.[105]

100 Botswana Mining Law of 1999, §§ 65 (3), 9.
101 Zambia Mines and Minerals Development Act (2015).
102 Namibia Mining Law of 1992.
103 Ethiopia Mining Law of 2010, § 61.
104 Ethiopia Mining Law of 2010, § 60.

Thus, a comprehensive and thorough approach to mine reclamation is consistent with the commitment of mining companies to the sustainable development of mineral resources. The certainty of a detailed approach to mine closure, including embedding reclamation at the very beginning of the mine planning process, allows a mining company to integrate reclamation in its assessment of project economics and assure that investors are fully informed of the full cost of a mining venture.

Thus, we can return to the key questions in designing an approach to decommissioning and reclamation, in light of the principles discussed above.

11.1 How can decommissioning and reclamation be incorporated into mine design and permitting?

We have seen that a thorough mine permitting regime includes decommissioning at the very beginning of the mine permitting process. The mine plan will include a description of the proposed post-mining land use, and will describe how the mine will proceed in a manner that allows proper decommissioning and reclamation. In short, the mine will be designed in a way that allows for its decommissioning and reclamation at the end of mine life, and permits issued for mine operations will include requirements designed to promote proper reclamation.

11.2 How can mine operations be conducted in a manner to prepare for decommissioning and reclamation?

These permitting requirements will set forth the steps that are required during mine operations to assure proper reclamation at the end of mine life. For example, as discussed above, the mine operator may be required to segregate and stockpile topsoil that is stripped from the site. Doing so will allow that topsoil to be replaced and reseeded to foster reclamation. The mine operator may also be required to manage water impacts in a manner that is protective of local fisheries or aquatic life. Doing so will promote the recovery of fisheries and aquatic biota at the end of mine operations. Where appropriate, mine operations can include contemporaneous or ongoing reclamation before the end of mine life. Through this process, the total disturbed area is reduced during mine life, making final reclamation more efficient and less expensive.

---

105 L Danielson (ed), *Mining, Minerals and Sustainable Development Report: Breaking New Ground* (International Institute for Environment and Development, 2002). See also Elizabeth Bastida and Tony Sanford, "Mine Closure in Latin America: A Review of Recent Developments in Argentina, Bolivia, Chile and Peru", www.sdsg.org/wp-content/uploads/2010/02/Mine-closure-Latin-America-Bastida-and-Sanford-for-SEMINA.pdf.

11.3 How can a regulator assure that a mining operation will complete decommissioning and reclamation at the end of mine life?

There are two key elements for regulators to address to assure complete decommissioning and reclamation at the end of mine life. First, as noted above, the mine plan needs to be designed to incorporate a reclamation plan as part of the front-end planning for the mine. This reclamation plan should include a cost estimate for reclamation. Second, the regulator will need access to the funds to complete mine reclamation if the mining company becomes insolvent or otherwise walks away from the project. As discussed above, those funds are often secured through a surety bond of some sort, but the security can come in many forms. Also, the form of security may depend on the jurisdiction where the mining operation occurs. In some instances, the mining company may be able to offer sufficient security based on its balance sheet, or the balance sheet of its parent or other affiliated company.

11.4 What test can be used to measure successful reclamation?

If the mining permit includes a reclamation plan as part of the permit process, then one of the key measures of successful reclamation will be whether the reclamation process as implemented comports with the reclamation plan. Overall, the regulator will look at the reclamation of surface lands, including any regrading and topsoiling, to assure that the reclaimed area is consistent with the required topography. The regulator will also look at geophysical stability, especially if the reclamation plan includes any highwalls or impoundments at the end of mine life. Finally, the regulator will retain jurisdiction over the site until revegetated areas have been adequately established.

11.5 At what point can a mine operator be relieved of any further responsibility for a reclaimed mine site?

In many cases, a mining company is relieved from regulatory oversight in stages, with portions of the operator's bond released upon the completion of benchmark events in the reclamation process, such as the completion of regrading and topsoiling, or the development of geotechnical stability improvements. In most cases, final bond release can be made where the operator shows that adequate vegetative cover has been established for a sufficient period. That period may vary depending on the climate where the mine is operating. There may be some lingering obligations, especially where the operator is required to run a water treatment facility after mine operations are complete.

Breaking New Ground ties successful reclamation to a forward-looking vision: "For a mine to contribute positively to sustainable development, closure objectives and impacts must be considered from project inception. The closure plan defines a vision of the end result of the process and sets concrete objectives to implement that vision. This forms an overall framework to guide all of the actions and decisions taken during the mine's life."[106] Only with some vision for the ending can a mining project properly begin its planning and operation. The regulatory approaches described in this chapter are designed to assure that the mining company seeks that vision, and provides a framework for capturing and implementing that vision.

106 *Breaking New Ground*, at 243.

About the authors

Felipe Allende
Senior associate, Bofill Mir & Álvarez Jana Abogados
fallende@bmaj.cl

Felipe Allende focuses his practice on natural resources, mining and energy law. He received his JD from *Universidad Diego Portales*, Santiago, Chile (2009) and was admitted to practise in Chile in 2010. During his studies, he participated in the Griffith University study abroad programme in Australia (2006). Mr Allende also holds an LLM in environmental and natural resources law and policy, specialising in mining and energy, from the University of Denver in Colorado, United States (2014), having been awarded the university's Natural Resources Scholarship.

Mr Allende has worked in Canada as an international associate in mining and securities at Cassels Brock & Blackwell (2014–15), and in Chile as counsel at Sonda SA (2010) and JACOBS Engineering (2010–11). He was highlighted as a leading lawyer in mining by the *Legal 500* in 2018.

Scot Anderson
Partner, Hogan Lovells
scot.anderson@hoganlovells.com

Scot Anderson's practice revolves around commercial transactions, regulatory advice and project development. Mr Anderson oversees complex transactions for natural resources clients, including the formation of new joint ventures, the divestiture of natural resources projects, and the acquisition of leases and reserves. He represents clients in administrative proceedings as well as federal and state courts. He regularly addresses issues related to mineral development of federal public lands and Indian lands, and also advises clients on international project development and transactions. He has worked on matters throughout North and South America, as well as Europe, Australia, Africa and Asia. Mr Anderson is the global head of the firm's energy and natural resources industry sector group.

David Budd
Partner, Cassels Brock & Blackwell LLP
dbudd@casselsbrock.com

David Budd is a partner at Cassels Brock & Blackwell LLP in both the Toronto and Vancouver offices. He is a member of the financial services group and the mining, equipment and asset financing and sports industry groups. As a member of the mining group, He has worked extensively with both lenders and resource companies in the mining sector on a variety of unsecured and senior secured debt facilities on both domestic and multi-jurisdictional transactions with an emphasis on project finance, acquisition and corporate credit facilities.

Mr Budd has extensive experience leading project financings in the mining sector across a number of jurisdictions, including Canada, the United States, Mexico and Colombia. In addition to his leadership in mining project finance, he is currently the deputy managing partner of Cassels Brock & Blackwell LLP and has achieved rankings in the *Canadian Legal Lexpert Director* (Asset Equipment Finance/ Leasing; Asset-Based Lending); *Who's Who Legal: Canada* (Sports & Entertainment) and *IFLR1000* (Notable Practitioner).

Luke Danielson
President, Sustainable Development Strategies Group
danielson@sdsg.org

Luke Danielson is an attorney, professor, researcher and consultant on minerals policy, national development strategies, and environmental and social performance in the mining and oil and gas industries. He is the president of a non-profit research institute, Sustainable Development Strategies Group.

Mr Danielson was executive director of the Mining Minerals and Sustainable Development Project at the International Institute for Environment and Development; the founding director of the Mining Policy Research Initiative, a project of the International Development Research Centre in Montevideo; and a professor at the University of Chile faculty of law.

He is known for his work on minerals policies at both international and national levels, and has worked with over a dozen governments, including those of Mozambique, Chile, Peru, Mongolia, Cuba and China. He was a 2015 inductee into the International Mining Technology Hall of Fame for his contributions to environmental management and stewardship.

Brian P Dominique
Partner, Cassels Brock & Blackwell LLP
bdominique@casselsbrock.com

Brian Dominique is a partner practising commercial law with an internationally recognised focus on the natural resources sector. He has acted on transactions involving mining projects across Canada (including the far north) and throughout the Americas, Africa, Australia, Europe and Asia, including advising clients that explore for, build and operate all manner of precious and base metal and other mineral projects. Mr Dominique has significant experience in mining exploration and mine development, production and metals/minerals sales, including alternative financing arrangements (farm-in agreements, joint ventures, royalty agreements, off-take agreements, streaming agreements), as well as all manner of operating agreements, easements, power purchase agreements, structuring joint ventures, partnerships and limited partnerships and acquisitions of related government rights. Mr Dominique also has extensive experience in Aboriginal law and negotiating with Indigenous peoples. He was called to the Ontario Bar in 1989.

Arend Hoekstra
Associate, Cassels Brock & Blackwell LLP
ahoekstra@casselsbrock.com

Arend Hoekstra is a practising lawyer and a chartered professional accountant. He has advised on mining projects, resource legislation and Aboriginal law issues, including the use of reserve lands and traditional lands, project-related environmental and regulatory matters, and Section 35 treaty negotiations. He acted as senior adviser to one of Canada's largest diamond mines and has served on the council of the Institute of Chartered Accountants of the Northwest Territories and Nunavut. Mr Hoekstra has a JD from the University of British Columbia and is a member of the law societies of British Columbia and the Northwest Territories (NWT).

Thomas Isaac
Partner, Cassels Brock & Blackwell LLP
tisaac@casselsbrock.com

Thomas Isaac has extensive national experience advising energy, oil, gas, pipeline, mining and forestry companies, lenders and investors, and federal, provincial, territorial and municipal governments on Aboriginal law. His published works on Aboriginal law have been cited with approval by Canadian courts, including the Supreme Court of Canada. Mr Isaac is a former chief treaty negotiator for the government of British Columbia and former assistant deputy minister responsible for establishing Nunavut for the government of the Northwest Territories (NWT). Mr Isaac was appointed special representative to the minister of Indigenous and northern affairs regarding a Section 35 Métis rights and reconciliation framework; served as the minister's special representative regarding claims negotiations in the Southeast NWT; and served as the minister's special representative to conduct exploratory discussions regarding the Gottfriedson class action lawsuit. Mr Isaac is a member of the law societies of British Columbia, Alberta, NWT, Nunavut and Yukon. He was named in 2018 by *Canadian Lawyer* as one of Canada's most influential lawyers.

Stuart MacGregor
Partner, Clayton Utz
smacgregor@claytonutz.com

Stuart MacGregor is a partner of Clayton Utz and the national industry group leader of the corporate energy and resources group. He is recognised as one of the world's leading energy and resources lawyers by International *Who's Who Legal, Chambers Guide, Best Lawyers, Doyle's Guide, Legal 500 Asia Pacific* and *IFLR1000*. He is known for his "profound technical ability" and "in-depth experience" in the energy and resources industry (*Who's Who Legal*). Mr MacGregor also lectures in mining and petroleum law at the University of Queensland. He has expertise in the full lifecycle of mining projects, from initial exploration to mine closure and rehabilitation. He advises on issues related to cross-border joint ventures, negotiation and drafting of commercial agreements, tender processes, long-term commodity agreements, commodity handling agreements and shipping agreements associated with major infrastructure projects, as well as local and cross-border acquisitions and disposals of projects and infrastructure.

Dawn G Meidinger
Director and shareholder, Fennemore Craig, PC
dmeidinger@fclaw.com

Dawn Meidinger been practising law for over 20 years, representing mining companies and land developers. She is chair of Fennemore Craig's natural resources, energy and environmental practice group, based in the Phoenix office. Previously, she worked as in-house counsel for Phelps Dodge Corporation (today Freeport-McMoRan Inc), where she specialised in mineral exploration, development and state and federal permitting on private and public lands. Her work continues in that specialty in addition to cultural resources and endangered species matters. Her expertise is bolstered by time spent in the private sector as the operational lead on the entitlement and development of residential and commercial real estate projects in Arizona.

Ms Meidinger is a graduate of the Arizona State University (1990) and the Sandra Day O'Connor College of Law (1996). She is a member of the Arizona State Bar.

Marie-Christine Moon
Lawyer, Clayton Utz
mmoon@claytonutz.com

Marie-Christine Moon graduated from Queensland University of Technology with a bachelor of laws (hons) and is currently a lawyer in the corporate energy and resource team at Clayton Utz. She has experience advising resources clients on corporate, commercial and regulatory issues associated with major resources projects, including joint ventures, native title, mine rehabilitation, local and cross-border acquisitions and disposals of projects. Ms Moon started her career as a judge's associate to a district court judge (also commissioned in the planning and environment court in Queensland) before commencing practice at Clayton Utz.

Patrick Petzall
Legal counsel, Trapeze Group
patrick.petzall@trapezegroup.com

Patrick Petzall is legal counsel for the Trapeze Group, an affiliate of Constellation Software, a Toronto Stock Exchange listed company. Previously, he was an associate with the construction group at Cassels Brock in Toronto and a consultant to Cassels Brock's business law/mining group; before that, he was a junior partner at one of Venezuela's leading law firms. He holds degrees in law, business and international affairs, and has been called to the bar in both Ontario and Venezuela.

He has extensive experience in negotiating, drafting, interpreting and managing risks of multimillion-dollar contracts in the areas of construction, oil and gas, mining, power generation and international project finance. Mr Petzall's experience includes drafting, negotiating and monitoring performance of large-scale engineering, procurement and construction contracts, public procurement, international project and public sector financing, as well as negotiating and drafting public-private partnership-related agreements.

Fred R Pletcher
Partner, Borden Ladner Gervais LLP
FPletcher@blg.com

Fred R Pletcher is a lawyer who works in the Vancouver office of the Canadian law firm Borden Ladner Gervais LLP. He is the chair of the firm's national mining group. He advises public companies, boards of directors and underwriters in connection with a broad range of transactions and issues, including mergers and acquisitions, corporate finance, corporate governance, shareholder activism, continuous disclosure and general commercial matters.

Although Mr Pletcher acts for clients across a diverse range of industries, he has a particular focus on mining and has worked in this area for over 25 years. He regularly acts for Canadian and international mining companies in connection with acquisitions and dispositions of mines, development projects and exploration properties; option agreements; joint venture agreements; project finance; royalty, streaming, off-take and ore sales agreements; confidentiality agreements; tenure issues; permitting; expropriations; and National Instrument 43-101 reporting obligations. Mr Pletcher frequently speaks and writes on mining law topics.

Darrell W Podowski
Partner, Cassels Brock & Blackwell LLP
dpodowski@casselsbrock.com

Darrell Podowski is a partner in the securities and mining groups of Cassels Brock & Blackwell LLP in Vancouver, Canada. He practises business and securities law, advising mining, industrial, entertainment, technology, and biotechnology companies on going public transactions, corporate finance, merger and acquisition transactions and general corporate commercial matters in Canada and internationally. Previously, he was corporate counsel with Teck Resources Limited, a major diversified mining company based in Vancouver, and prior to his law career he was an oil and gas exploration geophysicist with Dome Petroleum and Amoco Canada based in Alberta, Canada. As a result of his diversified experience, Mr Podowski possesses a particularly strong understanding of the practical commercial needs of his clients. In addition, he has particular expertise with Latin America-related transactions and has extensive offshore experience as a result of his two-and-a-half years working in Bermuda with a large offshore law firm. He holds a bachelor of science in geophysics and a bachelor of law, both from the University of British Columbia, Canada.

Jennifer Poirier
Lawyer, Cassels Brock & Blackwell LLP
jpoirier@casselsbrock.com

Jennifer Poirier is a lawyer in the securities group of Cassels Brock & Blackwell LLP in Vancouver, Canada. She has experience acting for issuers and dealers in connection with public offerings, private placements and mergers and acquisitions in Canada and globally in a variety of industries, including mining, renewable energy and technology. She also has experience advising on corporate commercial agreements, such as option agreements, joint venture agreements, memorandums of understanding, consulting agreements and partnership agreements.

Ms Poirier's recent transactions involve advising on various financings, mergers and acquisitions and other commercial transactions involving mining and renewable energy issuers. She is also an organiser of international not-for-profit mining association, the Young Mining Professionals, established to advance the global profile and leadership of the mining and exploration industry. Ms Poirier holds a bachelor of arts degree and a juris doctorate from the University of British Columbia, Canada.

Carla Potter
Lawyer, Cassels Brock & Blackwell LLP
cpotter@casselsbrock.com

Carla Potter is a lawyer in Cassels Brock & Blackwell LLP's financial services group. Her practice focuses on project financing, corporate financing and equipment finance matters, representing both lenders and borrowers involved in domestic and cross-border lending and leasing transactions and corporate reorganisations.

Ms Potter has been involved in several project finance transactions in the mining sector including as it relates to mining projects in Canada, Guyana, Mexico and Colombia, as well as several secured and unsecured corporate credit facilities for mining issuers. She is also a contributing author to texts regarding project finance in Canada.

Ms Potter is also a chartered professional accountant and chartered accountant with experience working at national accounting firms in both an assurance and business advisory capacity. She completed her JD degree from Queen's University and previously obtained an honours bachelor of commerce degree from Lakehead University.

Shea Small
Partner, McCarthy Tétrault
ssmall@mccarthy.ca

Shea Small is a corporate partner and co-leader of McCarthy Tétrault's global metals and mining group. Based in Toronto, Canada and London, United Kingdom, Mr Small has an international practice focused on M&A, capital markets, private equity and joint venture transactions. He has extensive experience in the metals and mining sector in Canada and around the globe, including acting for the major diversified mining companies, as well as private equity groups across a wide range of bulk, base metal, precious metal and speciality commodities.

Sarah A Strunk
Chair, director and shareholder,
Fennemore Craig, PC
sstrunk@fclaw.com

Sarah Strunk has spent more than 25 years representing mining and mineral related companies, as well as lenders to the mining industry. She began her association with the mining industry in 1992, when she joined the legal team at Cyprus Amax, as counsel to the copper/molybdenum division of the company. Her principal focus is on mergers and acquisitions, lending and sales transactions, with transaction experience all over the world. She is based in the Phoenix, Arizona office of Fennemore Craig.

Ms Strunk has a master's in law from New York University (Corporations) and is a graduate of the University of Kansas School of Law (1985). She is a member of the Bars of the states of Arizona, California, New York, Connecticut and Kansas.

Roger Taplin
Partner, McCarthy Tétrault
rtaplin@mccarthy.ca

Roger Taplin is a partner in McCarthy Tétrault's business law group and is co-leader of the global mining group. His practice is primarily focused on the areas of mergers and acquisitions and securities, particularly in the mining sector. Mr Taplin's experience includes advising on significant mining M&A transactions, business combinations, share and asset sales and joint ventures. He acts for issuers and underwriters on public and private securities offerings, as well as providing general securities regulatory and corporate advice to public companies. He has acted for mining clients and banks in relation to sizeable mining projects in numerous international jurisdictions.

Jordi G Ventura
President, Ventura & Associates LLC
jordigventura@gmail.com

Jordi Ventura is an international mining

attorney and Latin American specialist with vast experience advising junior and major US and Canadian mining companies throughout the Americas on social sustainability, sustainable development, corporate social responsibility and environmental matters.

Mr Ventura has provided clients with his expertise in diverse settings – as an independent consultant, at US and Canadian law firms, and as local general counsel for large multinational corporations in various countries in Latin America. In the mining industry, Mr Ventura has served as local general counsel for mining companies in Peru, Colombia, Mexico, Guatemala and Honduras. He has lived or worked in most countries in Latin America.

Mr Ventura is completely bilingual in Spanish and English on a professional level, fluent in Catalán and French, and conversational in Portuguese and Italian. He is also cross-cultural, possessing citizenship in both the United States and the European Union (a national of Spain).

Adam D Wanke
Counsel, McCarthy Tétrault
awanke@mccarthy.ca

Adam D Wanke is counsel in McCarthy Tétrault's business law group in the London, England office. His practice is primarily focused on corporate and commercial law, including mergers and acquisitions, particularly in the mining sector. Mr Wanke has experience advising clients in numerous international jurisdictions with respect to mining projects, mineral streaming transactions, joint ventures, mining royalties and asset and share sales.

About Globe Law and Business

Globe Law and Business was established in 2005, and from the very beginning, we set out to create law books which are sufficiently high level to be of real use to the experienced professional, yet still accessible and easy to navigate. Most of our authors are drawn from Magic Circle and other top commercial firms, both in the UK and internationally.

Our titles are carefully produced, with the utmost attention paid to editorial, design and production processes. We hope this results in high-quality books which are easy to read, and a pleasure to own. All our new books are also available as ebooks, which are compatible with most desktop, laptop and tablet devices.

We have recently expanded our portfolio to include a new range of journals, available both digitally and in hard copy format, and produced to the same high standards as our books.

We'd very much like to hear from you with your thoughts and ideas for improving what we offer. Please do feel free to email me on sian@globelawandbusiness.com with your views.

Sian O'Neill
Managing director
Globe Law and Business

www.globelawandbusiness.com

£140.00